SELF-TAUGHT
Navigation

SELF-TAUGHT
Navigation

Ten Easy Steps to Master Celestial Navigation

By Robert Y. Kittredge

Northland Publishing

First Edition
Seventh Printing 1989
ISBN 0-87358-496-1
Library of Congress Catalog Card Number 73-121015
Composed and Printed in the United States of America
3-89/3.5M/0226

*For all
the far-flung little ships
making wakes around the world.*

INCLUDED IN THIS BOOK
IS AN EXPLANATION OF
CAPTAIN KITTREDGE'S
SELF-DEVISED METHOD
OF HOLDING ANY COURSE
LINE WITH ONLY ONE
SIGHT EACH DAY.

CONTENTS

LIST OF ILLUSTRATIONS

PREFACE

YEARS AGO, during the Great Age of Sail, a life-and-death need did exist to cloud the simple process of celestial navigation with all the "hokum" and mystery possible. The desperate intention was to make it appear as a magic art, which only the special few could master. And there was a justifiable reason for this. The ever present threat of mutiny.

Because of the inhuman conditions men were forced to endure during those long voyages, the captain of each ship stood alone, pitted against his crew — the symbol and the cause of their privation. They, as well as he, knew how easily they could overpower him. The one deterrent the captain held to keep his crew at bay was that he and his First Mate, alone, could navigate. This First Mate, a fellow officer, in line for promotion to captaincy, could usually be relied upon to stand firm at the side of the captain. Thus, a fairly safe and even balance of forces was created. For any mutiny to be successful — as in the case of the Bounty — the rebellious crew realized that, of necessity, the First Mate had to be won over, if they were to have the slightest chance of steering the ship to some remote destination, in the hope of avoiding capture and hanging.

And so it became almost a ritual of survival for the captain and his mate to appear each day, at noon, on the high afterdeck of the ship, in full view of the crew below them, secretly and mysteriously handling the sextant between them, forever impressing the quarterdeck by every means at hand with the wonder, the difficulty, the intricate magic of navigation. They had the cabin boy swinging an hour-glass in strange circles on the end of a string. The captain positioned himself in certain exact relations to the ship before starting each sight. Then the mate had first to find the sun through a telescope, to verify certain auspicious aspects of its position. All this in the hope of deluding the crew into believing that no man among them ever could master what he was seeing done. For, had they but known — how simple it all was!

This was fine. Understandable. At least from the captain's point of view, all those many years ago. But, except in rare cases, on most yachts there is no threat of mutiny today. Yet it does seem, and this is substantiated by a great number of yachtsmen that I have met, that the several books presently in print tend to repeat this ancient tendency to cloud the simple process of celestial navigation with an aura of mystery.

I am not implying that the few books obtainable do not present a lot of valuable information. Yet I know of none of these books that, after several hundred pages of minor, secondary information, gives more than a few paragraphs at the very end, to any hint of what you do to actually navigate by celestial bodies. And there are none of them that explain, in detail, the one and only *simple* advanced method used by all air and sea ships since the last Great War.

Therefore, I address this book to the man who, I strongly feel, wants only to cut through all this wordy erudition and, with confidence, find that distant haven — to know exactly when it will appear over the horizon, and at what point on his bow. The intention of this small book is to convince that man how easy it is.

EDITOR'S NOTE

CELESTIAL NAVIGATION IS VALUABLE TO ANYONE traveling the wide seas, whatever the reasons or means. It is easy to learn, reliable, and inexpensive. It works every time. For the cost of a good sextant and a chronometer, a sailor can successfully navigate a ship to any port on the globe. As the author states in his "Helpful Hints" chapter: "Don't rely for safety on radios, direction finders, radar, Loran, Consul, etc. The only safe and sure way to know where you are is by continuous, thorough celestial navigation, backed up by careful, painstaking dead reckoning. Too often it is the gadget ship that is lost or in trouble."

The key to celestial navigation is consistency. Practice the steps contained in this book *every day* until they become second nature. Gain confidence in your abilities before you take to the open seas. If you have mechanized navigational systems, learn this one as backup.

Since the release of the first edition of *Self-Taught Navigation* in 1970, much has changed relative to navigation. Modern and state-of-the-art navigational systems have come into more popular and widespread use. Yet many cannot afford such expensive navigational systems; many are looking for a more reliable method; many are interested in the pure art of navigation; and still others wish to learn celestial navigation as a backup to their mechanized systems. For those people, we offer this book. There is no need to update or revise, because celestial navigation just hasn't changed—it is still the tried-and-true method it was during the Great Age of Sail.

No. 34   LINE OF POSITION USING H.O. 249   Area *MARTINIQUE*

Body	Date	D.R.Long.	D.R. Lat.	Log	WWV	Stop Watch	Local T.
SUN	*JUNE 23.76*	*62°02'W*	*15°09'N*	*487*	*15h 20m*	*03'.11s*	*11:23 Am*

Time

	h	m	s	
a	WWV	15	20	
b	Watch		03	11	- 02
d	GMT	15	23	09

Ass. Pos.-LHA (Almanac)

j	GHA Hours	44°29.3
l	Min.,& Sec.	5°47.3
p	(SHA)	
	(360)	360°
m	Total GHA	410°17.0
	Ass. Long.	62°17.
o	LHA	348 00

Sextant (Almanac)

f	Main	+15.7'	e
g	Dip	-2.5'	h
h	Total	+13.2'	i
e	Hs	75°36'	
h	Corr.	+13.0'	
i	Ho	75°49'	

Intercept-Azimuth (249)

q	Tabul. Hc	76°08'
r	d -34 Dec.Min. 26 0rr. -15
	True Hc	75°53
u	True Ho	75°49'
w	Intercept	4'
	T	A	✓
v	Z 53	= Zn 53°

Declination (Almanac)

a		
k	GMT Hour	N 23°26.0
n	"d" -0.0 Corr. -	0.0
s	True Dec.	N 23°26

Plotting

	Assumed Long.	62°17'
x	Zn 53°	Ass. Lat. 15°N
	Intercept	T	A 4

WORK SHEET
FOR SUN SIGHT
CHAPTER I

CHAPTER I

Ten Simple Steps
in Shooting the Sun

NAVIGATION BOOKS THAT I HAVE READ slowly work up, toward
the last chapter, to some vague hints as to the hazards you might
expect in attempting to shoot and plot a Line of Position — if they
hint about it at all.

So I have decided to reverse this process and tell you, as the
very first shot out of the gun, exactly what you so earnestly want
to know — tell you step by step what you do to take a Sun sight.
Then in the remainder of the book, I will attempt to clarify certain aspects of this process, and expand it into the Planets, the
Moon and Stars. I will extend it only to the point necessary for
you to understand *all* there is to know about what you are *doing*,
and why. So get your sextant and whatever watch you are going
to use — and let's get topside!

TO SHOOT THE SUN

1. Set up some type of accurate watch, exact to the second, on
Greenwich Time. The simplest way to do this is to turn on a
transistor radio of the type of the Zenith Trans-Oceanic to the
continuous night-and-day Time Signals, called **WWV**, sent out in
Greenwich Time from Ft. Collins, Colo., and obtainable on the
Zenith Trans-Oceanic anywhere in the world. (Found on Fre-
quency, 5.0, 10.0, and 15.0 MC; listen for the sound of a clock
ticking.) At the beginning of any full one minute interval, when
the exact Grennwich Time is given in voice, press a stop watch
to start it going. The announced time plus the accumulated time
on the Stop Watch at the instant of making the shot, will give
you the precise time of the shot. This is such an easy procedure,

there is no sense in paying twice the price of the radio for a cheap chronometer and then trying to remember to keep it wound every day. Write down this beginning **WWV** starting Time on your Work Sheet, so you won't forget it. See work sheet in this book. (Incidentally, if you are really going to sea, have a printer make you up several hundred Work Sheet Forms similar to the one shown here.) So we put down our **WWV** Time (a) on the Work Sheet.

2. Read the Taff Log, if you are trailing one (helpful, but not essential; any roughly close guess at your D.R. position will not affect the accuracy of your Fix.) If you are dragging a Log Spinner, write down your total accumulated miles (c).

3. Take your sextant, which we shall assume has been checked for accuracy in regard to horizon-horizon reading zero, and get topside somewhere so that you can brace yourself so both hands are free. I prefer standing in the companionway with just head and shoulders up, which is good, too, when it's wet, because you can hold the sextant down under the coachroof, if you have to wait for the sun to crawl out from behind a cloud. (Your sextant should be equipped with a safety lanyard to slip around your neck.)

With the stop watch running, and hung on a string so it rests on the center of your chest — or better held by someone else waiting for you to call "Mark," roughly make a guess at the Sun's altitude, (90° being straight overhead, halfway to the horizon being 45°) and pointing the sextant toward the Sun as you hold it to your eye, slowly move the arm back and forth, while slightly swinging the sextant from side to side, until you find the Sun in the eyepiece. If the image is too bright for your eye, alter the upper filters till it becomes just a nice clean fairly-dull ball; the lower filters are to dull any bright shine on the horizon. By the micrometer knob refinement device on the arm slowly bring the Sun up or down until, with your best effort at exactitude, you have its lower edge (called lower limb) just barely touching the horizon. The accuracy of this hair's breadth contact, or "kiss," is the accuracy of your sight, for each minute of a degree of arc on that knob represents a mile, one way or the other, in your ultimate Position Line.

At the instant when all seems the most exact you can make it, call "Mark," or quickly reach down and snap off the watch, in which case you should deduct about two seconds in the Time registered, for the over-run caused by this movement of your hand (b). A second or two variation in the accuracy of timing a sight is of small importance compared with the accuracy of the **Sun's** angle you are measuring. Carefully checking the angular reading in degrees on the curved main scale of the sextant and adding your minutes of arc from the micrometer vernier, write down this Apparent Sextant Height, or Hs (e). Also total in the Time Block on your Work Sheet the exact total Greenwich Time of the sight in hours, minutes and seconds (d), by adding the Stop Watch Time (b) to the beginning Time (a) taken on **WWV** (a & b). Replace the sextant gently in its box, if possible not disturbing the arm, so you can even check it again later. And now you are ready for a little Fourth Grade arithmetic, which, once you are in your stride, will not take more than five minutes.

4. Every sextant angle of every observation of every celestial body has to be corrected, by a combination of two corrections. (I'll explain why later in this book; for the moment, just take my word for it. Like dialing that long distance phone number, you've got to start with a "one," or your call won't get anywhere.) Open the front cover of your **Nautical Almanac,** current for the year (purchased in just about any Marine Supply & Chart Store) and you will find three columns. **Sun. Stars and Planets. Dip.** Mark them at the top, one, two and three.

You will notice the **Sun** column is divided into two columns, cutting the year in half, or two six-month periods. Select the correct half year, depending on what month it is, and run down the left hand side under the heading Apparent Altitude, until you come to the nearest exact or larger angle, corresponding to the angle you just have measured from the **sun,** in this bold-face left column of degrees and minutes. You will see that these bold-face figures always fall between two lighter figures in the next closest column, under the heading Lower Limb; always take the uppermost in position — the lesser in amount — of these two choices, noting that this is in **minutes** of arc (not degrees) and that for the **sun** it is always **plus,** or **added** to the Hs Apparent Sextant Angle you have just taken directly from the **sun** (f). But don't add it yet; we have another correction to combine with it.

We are, for this sight, not interested in the **Stars and Planets** column. But every sight is concerned with **Dip,** or Height of Eye Correction, in the third, or most right hand column, inside this same **front cover.** This is because for each ship, and for your position of observation on it, the height of your eye above the water naturally varies, from ship to ship, pertaining to any sight.

Consequently, all your Sight Reduction Tables, for simplification, are based on water level. Therefore, as accurately as possible (and you need only do this once, for the correction will be the same for all sights, if you take them all from the same place) determine the height of your eye above the water in feet. Find the nearest equivalent to this in the most left hand column, No. 3, under **Dip,** again taking the uppermost of the two corresponding figures in the Correction Column, noting that this is always a **minus** amount (g). So the easy way to handle this is first to combine this **Dip** correction with your first, or **Main, No. 1, Sun** correction, and then add the remainder (h) to your Hs angle (e). For example, following the Work Sheet given in this book, let us say your Hs was 75° 36′; month of June; your **main** correction would be plus 15.7′. Then let us say your **Dip** is 6.5 feet, as is my ship; your **Dip** correction would always be minus 2.5′. Combining these we get **plus** 13.2′, which we then add to our Hs of 75° 36′, for a Corrected Sextant Height, or **observed height,** or Ho, of 75° 49′ (i). All the above should be put down on your Work Sheet, for the record and for checking. That 13.2 minutes of correction that you added to your Hs, will make a difference of thirteen (rounded-off) miles in your Line of Position, so it is important. And that is all there is to correcting the Apparent Sextant Angle, which must be done for each and every sight. In time, this should take you about 30 seconds to accomplish.

5. Now, and don't panic — this is about as simple as looking up a number in the phone book — we turn to one exact page in the body of **white pages** of the same Nautical Almanac, which you will discover is nothing but a large calendar in these White Pages. So turn to the page in June for whatever the date may be, as on our Work Sheet for the 23rd, 1970, which you will find running across Two pages, with two other days occupying the same pages.

At first glance all this information may look confusing like a page in the telephone book, but it couldn't be simpler. On the left hand page is data pertaining to the **Stars and Planets,** which we are not interested in for the moment. On the right hand page, the first column at the far left, under the heading **Sun,** is what we are looking for. You will notice it starts with 00 hours (midnight) and descends to 23 hours (11 p.m.) which is not your Time on your ship, but the Time for Greenwich, England, as indicated by the G.M.T. at the top of this hours column. (In no time at all, this business of everything being in Greenwich Time will fall into place as logically as your knowing that if you place a long distance call at 8 p.m. in California to New York, you will wake Grandma up, because it will be 11 p.m. there.)

Under the **Sun** heading, to the right of this hours column you will find two columns: **GHA** (Greenwich Hour Angle); and **Dec.** (Declination). And here we are going to have our first little explanation of what was going with the **Sun** at the moment you hollered "Mark." You will remember I promised to try to hold this book to just what was necessary to get your ship from here to there. Well, it is very important that you know what this **GHA** and **Declination** means.

Essentially, both these terms are exactly, for a celestial body, such as the **sun,** what Longitude and Latitude are for your ship. **GHA** is an angular measurement pivoted on the North Pole and swinging westward from Greenwich, an ever-widening angle (through each 24 hours) until at 360° it is exactly back at Greenwich, much like the continuous advancing shadow of sunrise swinging daily around the earth must appear to North Star. To be able to measure any point on the earth, a basis of measurement had to start somewhere, and so some ancient brain just decided on Greenwich as "zero," so you can bet he was an Englishman, which simply means that when it is noon there, with the **sun** straight overhead, the **sun's GHA** is then "zero" (or very roughly so, as **Sun Time** and Clock Time are always close, but rarely identical). So you can check what I've just said about 12 GMT equalling 0° GHA in this **Sun** column anywhere in your Almanac, remembering of course that 360° and 0° are the same.

So, starting at noon at 0° over Greenwich, and traveling at 900 nautical miles an hour at the Equator (I know, it is really our Earth turning) the **Sun** will have traveled 21,600 nautical miles or 360° and by noon the next day it will be back over Greenwich again. Therefore, all those two columns (**GMT** and **GHA**) are is a Time Table of the Sun's Longitude around the Earth given in hour intervals for that day in Greenwich Time. But please don't worry about anything I've just said.

To work our sight what you are first interested in on this white page is your **WWV** full hour Time, which was given to you on the radio in Greenwich Time, and its **G.H.A.** equivalent, exactly as if looking up the Time of Arrival of a train in a Time Table (j). And the same goes for **Dec.** (k) which, just take my word for it, is the Sun's Latitude for that Hour of that day of the year. Simply copy them both down as they are from your Almanac onto the proper lines of your Work Sheet, being sure to give your **Dec.** the designation **N** or **S**, as shown.

I'm going to say this all in one other way, if you'll bear with me. Think of your sextant as a gun, and the **Sun** as a very fast flying bird, and your shouting "Mark" as being the instant you pulled the trigger. Assume you hit the bird and it fell at that exact second directly downward toward the center of the Earth, as if pulled by instantaneous gigantic gravity. Then the spot, or "splash," where it hit the Earth's crust, known as the **GP** (Geographic Position) of the celestial body at that instant, is what we are interested in right now — a point to measure our own ship's position from. So we write down those two figures on our Work Sheet (j and k) to enable us to combine a few other things with them.

You must understand here, and this again is important, that those two figures you have just written down from the **white** calendar page represent the **sun's** position for only the full hour of your total time. Unless by some rare fluke your instant of "Mark" hit exactly on the nose of a full hour, we still have the minutes and seconds amount of your WWV and Stop Watch total time to include with this hour amount, to arrive at the exact Time that bird, the **sun,** hit the water at GP. (Note: for the moment glance down at the very foot of the white page under the **Sun** column. There is a little S.D., which you can forget. There is also a very small "d" value given. Write it down on your Work Sheet (n) **plus** if the Declinations in the column are increasing, **minus** if they are decreasing. I'll mention more about "d" in a moment.)

So here we go after the GHA value for those minutes and seconds, for another ten second maneuver.

Turn to the **yellow pages** (not the phone book) in your Almanac, headed **Increments and Corrections.** Find the proper half page designated in minutes (23^m) in the upper corners in heavy numerals, progressing from 0 to 59. For each **minute** you will find six columns. At this moment you are only interested in the **Sun and Planets** column. The very left hand column is headed by a little "s" meaning **seconds.** Run down that column to the amount of **seconds** you have on your total WWV Time for your sight, as read from your Stop Watch at the moment of "Mark." On this Work Sheet we have 9 seconds. Note that these increment figures given are in degrees and minutes of arc. Add this

Minutes and Seconds Increment (l) in correct alignment to the **GHA Full Hour Amount,** which you wrote down from the White Page (j). This is now your total G.H.A. (m) (essentially Longitude) for the precise moment of "Mark," or "splash," or GP of the Sun for this sight.

Now we must do the same thing for the **Dec.,** or declination, or Latitude, of the **Sun,** and we will have its full GP all sewed up. Let us assume that little "d" amount at the bottom of the Sun column on the **white** page was 0.0, as in the case of our Work Sheet. While you are still on that **same** page in the six columns of the **minutes** page in the **yellow** pages, where you just took the amount of additional GHA for the minutes and seconds of this sight, the last three of these columns are headed "v or 'd' correction." This is always a very slight correction, compared to the usual Minutes and Seconds correction, the "d" pertaining to the Declination change during those same minutes and seconds. All it is is the infinitesimal amount of increase or decrease in latitude (Declination) which the **Sun** made between the full hour amount given on the **white** page, and your minutes and seconds, because the **Sun** advances or recedes very slightly toward or away from the Equator each hour of the year in its semi-annual season-making march across the Equator. But it is always good to include it, though granted it's a bore.

So, to accomplish this we go down the fourth row there, under "v or d" correction of the **same** minute section, looking in those very small figures for 0.0, and beside it, under the correction column, we find 0.0 (n). We are lucky. The amount of our minutes of time was so small, there is no correction. Fine. But if there had been, turn back to the **white** page; look again at the **Dec.** column for the day of this sight. Is the amount of Declination **increasing,** as the column goes down, or **decreasing?** If increasing, add this fussy little amount to the Declination; if decreasing, subtract it. That's **all** for the Almanac. Hand it down to your wife, though in all probability she's asleep by now, so confident is she in your ability to navigate.

6. We're coming right along! We'll have our Line of Position plotted on the chart in jig time, now!

What we have done so far all relates to the **Sun** and its **GP.** We haven't done much of anything regarding the whereabouts of our small ship, which is what we are really concerned about. But again, we can't ever measure something unless we first have some beginning point of reference to measure from, which we now do

have — the **Sun's GP.** So now we must begin to combine all that
we do know about our ship with this GP of the Sun. Well, what
do we know?

All that we have to work with out there, in that vastness of
water is the Sextant angle that we took, which is a very accurate
measure (we hope) and our **Dead Reckoning Latitude and Longitude.** This D.R. position, naturally, is always somewhat fuzzy,
due to drift, current and a lot of guess, no matter how accurately
we try to keep it. But this dead reckoning Longitude and Latitude is always quite sufficient, even if we are off by miles and
miles (your intercept is just that much greater, which we will
come to in a moment.) We really only use this D.R., or guess at
our position, to set up what we call an **assumed position.** And
this is perhaps the most important part of any sight and, for some
reason the most confusing, simple as it is, and therefore needs
some explanation. But again, this is so logical, there is nothing at
all to be frightened about, that is — if you got through the
Fourth Grade.

Let's begin with a couple of simple examples.

Suppose you are standing in a dark room, blindfolded, but you
know the furniture arrangement of the room, and so do I. How
can I tell you exactly where you are? I say to you, let's "assume"
you are standing at the corner of the table, which is nearest to
the hallway door, would you know where you were? "Yes," you
say. "Well," I answer, "you're not quite there. You are exactly two
and a half feet from that table corner in the direction of the
corner window." Now, do you know where you are? You know
precisely.

How could we have done it without assuming some point of
reference?

To take another example, which is a little more like a sea chart
and a ship somewhere on it. Imagine a large window of many
small similar steel-framed panes, and there is a fly sitting at some
point on one of those panes. Then let us imagine that at each one
of the many junctions, or crosses, of these frames we have pasted
a 360° protractor. You are in the other room with a small drawing
of the window to exact proportions. How can I best tell you
where the fly is? Sure. You're way ahead of me.

I say, "Assume he's at the junction of three panes down and
four over from the right. Well, he's not quite there —. Reading
from the protractor at that point, he's two and a half inches from
it in the direction of 62°, or 247°. Or I could say, slide that pro-

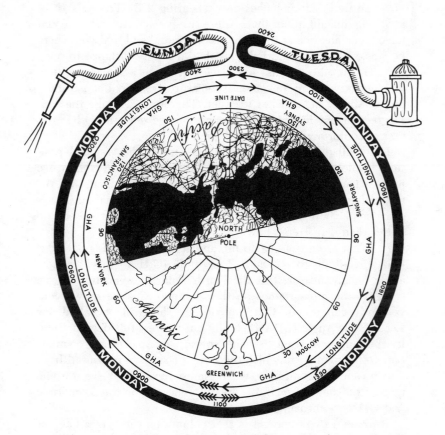

TIME & LONGITUDE AROUND THE WORLD

It is eleven in the morning at Greenwich, England. In exactly one hour it will be noon at Greenwich, GHA 0° for the sun. Just then, in the space of three seconds at the Date Line (180° East or West Longitude) it will have been Sunday, Monday and Tuesday; that is why you always jump from one day to another at this point.

tractor to the left exactly fifteen inches from that intersection, and now the fly lies four inches away, bearing 233° True." Is there anything difficult about that? There surely isn't. Actually, I could have taken any one of those intersections and still told you where he was, only we would have just had a longer distance to measure. Or, in other words, **any place can be an assumed position,** as long as we can conveniently measure from it, but naturally one relatively near makes it easier. Therefore, we take the Full Degrees of our D.R. Longitude and change only the D.R. Minutes to conform with the GHA Minutes.

That's all there is to the great mystery of the **assumed position.** In plotting a celestial Line of Position (LOP) on a chart, we just can't work without it. The oceans are just too damn big. We would have at least just to throw a coin indiscriminately on the chart somewhere and say, "let's relate it to that." But we can do better than that. Here's how.

7.   What we are going to do, and it's very easy, is relate our D.R. Longitude to the Sun's Longitude (G.H.A.) to get a new angle (again measured at the North Pole) in **some full degree** (roughly the cross frames on the window) as close as we conveniently can to our D.R. Longitude. This will be called our **Local Hour Angle (LHA),** actually our nearest **full degree** angular measurement from the **sun** "splash," in exact lesser counterpart to the manner in which the **sun** "splash" was measured from Greenwich. Hang on now, we're almost through! We've got one more little knot to iron out — the difference in the way **Longitude** and **GHA** are measured.

The **Sun,** as you noticed in that column on the White Page of the Almanac sweeps right on, around and around the earth each day, from East to West — it never goes any other way. But whoever set up the charts of the world didn't follow that pattern, he dreamed up a different idea. Chart **Longitude** begins at zero on the Greenwich meridian and **goes both ways,** East and West, each halfway around the world for 180°, finally coming together in the middle of the Pacific Ocean on the backside of our little planet — in fact, probably the best place for this mess to happen, as we shall see.

The reason for doing this **two way** Longitude gimmick was to try to keep the days straight — at least on the side of the globe where the Englishman lived. Thus, if he had to go around the world, he did all his adjusting on the back side, where it didn't cause too much disruption in London. Because, you see, when it's noon on Monday in Greenwich, it was Sunday a minute before in Midway and it's Tuesday in Beringovski in Russia, so Intelligent people didn't want to get back to England thinking it was Saturday, when it was really Sunday — but we don't want to get mixed up in all this; by the time you get around to the Date-Line of East-West 180° Longitude, it will all have fallen into place like a soaked Skipper in a wetter bunk.

For the moment, just take my word for this, and fix it firmly forever in the back of your head. When in **West** Longitude (all American waters) you **subtract** your ship's **D.R. Longitude** from the GHA of every sight. When in **East** Longitude (Indian Ocean, Fiji) you **add** your ship's **D.R. Longitude** to the GHA of each sight.

So, we will do this now, very simply, as shown on our Work Sheet, subtracting our **West** Longitude (arrow) from our total GHA (m) to arrive at our **Local Hour Angle (LHA)** (o). But how can we make this always come out in a **full degree,** because our next Books are all set up only in full degrees? It couldn't be easier. Invariably, your total GHA will be composed of **degrees** and **minutes.** All you do, when you are in **West** Longitude and **subtracting,** is take your D.R. Longitude and **give it the same minutes as the GHA minutes,** so that the subtraction wipes out the minutes. In **East** Longitude, when you are **adding,** you change the **minutes** of your D.R. Longitude to equal the **difference** between your **GHA Minutes** and sixty (60), so that in your addition they always amount to sixty minutes, which wipes them out into a **full degree.** If we didn't get our LHA into a **full degree,** our next book of tables would be so heavy we couldn't lift it.

And so here, in the process of arriving at your **Local Hour Angle** (which is what you need in working up each sight to get into your 249 Book), you arrive also automatically at your **Assumed Longitude,** the dictated part of your **Assumed Position,** for you merely select your **Assumed Latitude** from the nearest **full degree** of Latitude to your D.R. Position. And there you are. Your **Assumed Position.** And you arrive at it in the same way for every other sight, be it **Sun, Stars, Planet, Moon.** The moment you do your subtraction or addition, combining your D.R. Longitude in compatible **minutes** with your GHA of any celestial body, substituting the necessary minutes onto your Longitude to make it all come out to a **full degree** of LHA, there automatically you create your **Assumed Longitude, which you plot as**

a point on your full degree of Assumed Latitude. So, once again, your **LHA,** and your **Assumed Latitude** are always in **full degrees,** without **minutes.** Not very difficult, is it?

Incidentally, if you ever come to make the above subtraction and find your GHA is smaller than your ship's Longitude, just **add** 360° to the GHA (p) and go right on and forget it. You never have to give it back. I don't know how this works, but it does — something to do with making a full circle and arriving back where you were. Maybe you can figure it out (m). Also, when in East Longitude, if the resulting LHA is greater than 360°, subtract 360°. You rarely have to do this, but I have used a situation on our Work Sheet, just so you'll know.

8. The **Local Hour Angle (LHA),** which you now have, is always the one imponderable you have got to arrive at, to get into your next book, Volume II, of the 249 Sight Reduction Tables. And now let me give you a word of advice.

Don't ever, please, let anybody, at this precious moment of your starting to navigate, influence you to buy any other tables but the **249 Sight Reduction Tables for Air Navigation.** I know, it's a boat — not an airplane, that we are going to use them on. All other methods are merely hangovers of the Dark Ages of navigation and should be outlawed by an Act of Congress. This, among many others, includes 208, 211, 214, Grandma Moses' Condensed Tables for Whist and Certain Nautical Problems, and the whole lot of little thumb-worn Miracle Publications that have found their way out of old sea chests into the trembling eraser-blunted fingers of the well-meaning. Trouble is, every old "dog's body" who ever ran away to sea and picked up navigation before the turn of the century is **absolutely** convinced that his fine old tried-and-proven method is the **only one** — and for him it always will be! Because you'll never convince him that he can do more accurately in five minutes, what now takes him three hours and a dozen volumes and nine sheets of paper and a book of logarithms and a Chinese Abacus, to finally sweat it out. Just smile and make a wide circle around him on your way into the store to buy your **249 Sight Reduction Tables for Air Navigation.**

It comes in **three** volumes. For all Trade Wind cruising, Vol. II is basically enough. Volume III is for Latitudes above Latitude 40°. Brrrrrrr! **Volume I, Selected Stars,** is a special bit of magic which, I promise you, though essentially you don't need it, you will come to swear by. I will explain it in a separate chapter later in this book.

Now that we've gotten squared away on the **one and only** method to use, the **U.S. Navy Publication No. 249,** let's see how simple it is to use it. We have now worked up our sight to the point that we have all the material we need to enter. We open this book, Vol. II, to the proper Latitude page designated by our **Assumed Latitude.**

You will notice that the first page in this book is Lat. 0°; the last page is Lat. 39°. So the area covered is anywhere in the world on either side of the Equator, to 39° 59′ North or 39° 59′ South, which covers just about any circumnavigation of the globe. Except for 0°, there are six pages of Tables for each **full degree** of Latitude. You will notice also that each full degree of Latitude, over six pages, is broken down into columns of **full degrees of Declination** from 0° to 30°. The Latitude degrees apply to the chart; the Declination degrees apply to the celestial body.

Each group of six pages is further broken down into **Declination** of the **same** name as Latitude (N & N; S & S); and **Declination Contrary** name to Latitude (N & S; S & N). This simply means that if you are North or South of the Equator (if you've been doing any sort of navigation, you should know which side you're on by now) for half of each year you will have the **Sun** on your side (the Dec. for any date for all celestial bodies is always given in the Almanac) and for the rest of the year it will be on the other. So, as you change back and forth over the Line, it should be fairly simple to figure this out.

Thus, going back to our Work Sheet, if our **Assumed Latitude** is 15N and the date is June 23rd, the **Sun** is very definitely in North Declination (about as far North as it can get) so we turn to the one page in the 249 Book which will give us Lat. 15°, **Dec. same** as Lat., and Dec. 23°, which naturally has to be found in the Dec. area of 15° to 29° of Declination. (Just in case there's any possible confusion, this happens to be page 93.)

And now, if you look around, you will see there is an LHA column on each side of each page, and the beauty of this is, if you don't find the LHA you are looking for on one side, you can usually always find it on the other side. Incidentally, if you don't find it, you made some mistake, usually some stupid little goof in your addition or substraction — so right there, go back and find what is wrong. Let us presume we are looking for LHA 348. We find it right near the top of the page, on the right hand side.

Now, if we carefully carry that 348 line across to the Dec. column of 23°, look at all the magic we find there! All this amounts to is that someone else has done for us what used to take hours of pencil pushing by the navigator, and still does for those "sea-chest lads," so this unknown egg-head has saved us one hell of a lot of hard work. And here, right at this point, is something **important** and **helpful:** an almost positive proof that your work so far has been correct and accurate is that, the figure under this Hc column that we just have come to (q), should be very close to our Corrected Sextant Angle (i) worked up a while back from the Almanac, as the first step in this sight. And, it does! Which is always encouraging.

And here again, if this comparison between your Ho and tabulated Hc (Tabulated in the 249 Book) is way off, more than 40′ or 50′, you had better stop there and go back and start checking. And as I've just said, invariably you will find that almost every error you will ever make (except confusing one Star for another in your sextant) will be some careless little stupidity, because you are tired, or have allowed yourself that one extra drink. For, if given half a chance, your sextant and these books do not make mistakes. Check your sextant angle again, on the sextant itself; if you had a large amount of minutes of arc on the vernier, you can very easily make the common mistake of taking the full degree higher than the correct one from the curved Main Scale of the sextant. Or you have copied the wrong GHA for your hours in the Almanac. Or in your simple adding or subtracting you merely forgot to carry the **one.** Find where you goofed and everything will fall into place. Perhaps you forgot to correct your sextant angle? If you can't find it quickly, shoot the shot over.

But now that the Hc in the 249 Book **does** tally fairly close with our Ho (Corrected Sextant Angle), copy those three numbers down on your Work Sheet from the 23° Dec. column of your 249 Book (8 q, r), noting carefully whether the second number, "d" is **plus** or **minus.** This is **not** the same "d" we took rather casually in the Almanac. The last figure, the Z, pertains to angular direction, or Azimuth, of the celestial body from your ship at the moment of "Mark," or Splash," which we will use in a few moments for plotting our actual position from our **assumed position,** on the chart. All these Z or Azimuths are not yet **true** 360° bearings. (And remember you **never** put any bearings of any kind on any chart, except **true** bearings!)

9. We're really coming into the home stretch. Hang on!

With our increments for minutes and seconds from the Yellow

pages, we have already completed and satisfied the full amount of **GHA** for our **GMT** total time.

By the same token, we now have to take care of the **minutes of Declination**, for, as you see, the figures given on this Latitude page of the 249 Book are for **full degrees** of Declination only. So there is yet one more little step we have to go through, but it's about the easiest so far.

Take the Minutes of True Declination (s), written down on your Work Sheet from the Almanac (when you took the GHA hours) way back in the beginning. We are now going to combine these Minutes of Declination with this **plus** or **minus** number, given under the "d" column on the Latitude page of the 249 Book. Please turn to the very **last** page in this book, **Table 5.**

In this **Table 5,** it makes no difference which of these two numbers you apply crossways, and which you apply up and down — it works either way. Any way you slice, just find the little **correction** number, out in that wilderness of small numbers, that corresponds to the crossing of your **Minutes** of Declination and your "d" value, and giving it the proper **plus** or **minus,** combine it with the Tabulated **Minutes** of the Hc (Computed Height) (q) given in the **Dec.** column on the Latitude Page, to give you your **Total** or **True Hc** (u), or your **Computed Height,** which you will now compare with your Ho, or **Corrected Sextant Height.**

And what does this mean?

This True Hc (Computed Height) simply means that if you really were exactly at your **Assumed Position,** which is what you have fed into this book, your sextant reading (and the book knows what it is talking about) would have been this Hc angle, regardless of what you measured with your sextant.

We are back to the fly on the window!

Where you really are (for you were pretty sure all this time you weren't at that **Assumed Position**) is the **difference** in miles between this True Hc (**computed height,** which pertains exactly and **only** to the **Assumed Position**) and the Corrected Sextant Angle (Ho) you actually made!

Well, you may have known it all along, but a **minute** (') of Sextant arc happens to equal precisely **one** nautical **mile**! So aren't we lucky?

But in **what direction** do these miles bear from our **Assumed Position**? You are right. That "Z" (v, on your Work Sheet) tells you just this. Then let us say there is a **difference** in minutes, or miles, known as **intercept** (w) of four **minutes,** or four **miles, and** with our **parallel** rule we transfer our "Z," or **azimuth** from the compass rose on the chart over to our **Assumed Position** point, and striking it off through that point going both ways, how do we know whether this **intercept** is **toward** the Sun from our **Assumed Position,** or away from the Sun from our **Assumed Position?**

Easy! Just always remember this! GOAT!

If the Ho is the larger angle (Corrected Sextant Angle) your **intercept** is **toward** the body. Greater Observed Angle Toward.

If the Hc is the larger angle (Computed Angle, as in our case) your **intercept** is **away** from the body.

(You can remember it by, "T" (Toward) and "O" as being close together in the alphabet; and "a" (Away) and "c" as being close together.) Or better, think it out — that the nearer to the Sun you move, the more the angle widens, whereas, the farther away from it you move the more the angle narrows.

So, all we have left, before we plot, is a little word on how to use that "Z," in relation to the Compass Rose. You will notice that the "Z" Azimuth bearing never goes above 180° on the columns of the Latitude Pages, yet we have 360 possible degrees of arc that the Sun could bear from us.

Look at the **upper** and **lower** left hand corner of every page of this 249 Book, and you will see a little Equation under the heading N. Lat. and S. Lat. "Zn" (x) simply means the "Z" transposed into the **full** arc, of 360° of bearing. Half the time, as you will notice in the Equation, "Zn" already equals "Z." The rest of the time to get "Zn," you merely have to subtract "Z" from 360°. So if your "Z" is 120° and your LHA is less than 180°, if you are in **north** Latitude, the Azimuth that you will move from your compass rose on the chart to your **Assumed Position** will be 240°.

So now to plotting. (I hope it's not a plot to assassinate the author.) But first it might be a good idea, if you haven't done so already, to check out every step we have done so far, as shown on the Work Sheet of an actual sight, which appears at the beginning of this chapter. Just to see if everything is as clear as a fifth magnitude star on a cloudy night.

10.   **Plotting your Line of Position (LOP) on the Chart.**

We should take time out here, long enough to enumerate the few basic tools, which you must have to navigate. Naturally, you will have a sextant, better two, and some type of accurate watch, as well as a couple of Stop Watches and a radio capable of bringing in WWV anywhere in the world (Zenith Trans-Oceanic). Beside these, you must also have a pair of dividers, of which the one-hand type are quite the best; a strong pair of parallel rules, for "walking" azimuths and course lines around the chart from various roses; you need a right triangle, that you can see through. A pencil sharpener, a couple of boxes of good wooden pencils of medium hardness, and a high quality large eraser. These are the basic tools. There are numerous other expensive and complicated gadgets, which you can buy to clutter up the few drawers you have on board. Forget them.

In plotting our sight from our Work Sheet, the best and most helpful procedure I have found is to have a pad or two of small Navy type plotting sheets, which can be purchased in any chart house, in size a trifle larger than a sheet of typewriting paper, each with a 360° rose in the center, and the Latitude lines fixed and printed. In the lower right hand corner of each sheet is a very simple graph-type scale on which, depending on your Latitude as indicated, you measure and draw in your own correctly proportioned Longitude lines, for that area of water you are crossing.

These small plotting sheets are handy, easy to work on and very accurate. Two sheets (for you use both sides) will get you across any ocean. As each group of sights is worked up into a **fix**, you can transfer them from this sheet to your ocean chart, as often as you choose, each noon, or more frequently, as you wish, using your dividers. Thus you keep your charts neat and clean. If there is anything cumbersome to work on at sea, it's a chart — even if you waste half your main cabin space on a chart table. As to chart tables, I think one hinging up to a bulkhead, lowering down to rest on the cabin table, is the best solution and saves all that space. But let's get going with our plotting.

On our plotting sheet we have marked and ruled off our Longitude lines in the proper relation to our Latitude. Let's number our four or five Latitude parallels and Longitude meridians. Measure off from the chart with your dividers our point of departure and carefully transfer the beginning portion of our course line, as much as will fit, onto our plotting sheet. Naturally at sea, we will always be working from the last **fix**, using our new Log Distance run advanced on our Course Line, to set up our new Dead Reckoning Position for determining each new **Assumed Position.**

But now we are going to plot our Sight No. 34, that we have been working up together, which appears at the beginning of this chapter, ready for plotting. And again you'll be surprised how easy it is.

Except for divisions of Longitude, as Long. 15° 30'; Long. 29° 42', all nautical measurements of **miles,** as you probably know (due to the distortion of Latitudes in the Mercator Projection) are taken from the up-and-down, or vertical edges, of every chart, where the **minutes** or **miles** are always accurate to the Earth. When you are right down close to the Equator, 15° North or South, this distortion is negligible, but we still from habit measure any distance in **miles** on the vertical scale.

So our first step in plotting our **Assumed Position** (62° 17' W; Lat. 15° N) is to find the intersection of Lat. 15° and Long. 62°. With our dividers measuring off 17' on the proper line in the corner scale, we place one leg of our dividers on this intersection just mentioned and swinging the other leg Westward along the Latitude line on our plotting sheet, we make a nice clean little pencil dot, representing 62° 17' W on Lat. 15° N.

The **Sun,** in the case of our Work Sheet, happened to be bearing NE of us at the time of our sight. (One usually thinks of the **Sun** rising in the **East** and swinging an arc into the South; but in this case it crossed to the North of us. Why? Because its Declination is so high and our ship is so near the Equator, our Latitude being considerably less than the Sun's Declination.) As I have explained and as the Work Sheet indicates, our ship was lying, at the time of the sight, four miles **away** from our **Assumed Position,** with the **Sun** bearing true 53°.

Set one edge of your parallel rule on 53° of the compass rose. Carefully walk it across to the dot of our **Assumed Position** and draw a light dashed line through the dot sufficiently long to accommodate your transparent triangle. Again using your divider, measure off four **minutes,** or miles, on the vertical scale of the plotting sheet. With one leg on the **Assumed Position,** swing the other **away** from the direction of the **Sun** on our **Azimuth** (dashed line) and make another dot. Grab your transparent triangle and strike a line through this second dot perpendicular to, or at right angles to, that **azimuth** line. That is your **line of position** (LOP), along which somewhere your small ship lies!

Why aren't we right on that dot?

We could be. Maybe we are, or close to it. All the sextant angle can ever tell us is, that we were just a certain **distance** in **miles** away from the **Sun** at the instant of "Mark"; as if we had been holding one end of a million mile long measuring tape, with the far end fastened to the **Sun,** actually so far that the swing of that tape would appear as a **straight line** across any portion of our little earth. In other words, if we can imagine a thousand boats all strung out along our **line of position,** which we just have drawn, they would **all** have measured the **same** sextant angle, at our **same** instant of "Mark." Granted, they would each have had a different **Assumed Position,** and different Work Sheets, but if they had been anywhere on this particular **line of position,** the result of **all** our sights at that instant would have been the same.

So what does this mean?

It means we **always** need **another** sight, at least one more, preferably several, from the **Sun,** or the **Moon,** or a **Star** (or even from a lighthouse or a headland), before we then can say — **where two position lines cross** — here we are! So then, after each of those imagined boats on our present **position line** has its **second** sight — each with different **intersecting lines** — each will know precisely where it is.

So what do we do, to find exactly where we are? We have only one **Sun.** It's a long time to wait till the first **Star** comes out. No, we don't have to wait that long. But we do have to wait. **Long enough for the sun to move appreciably,** and so give us a **different azimuth,** and the wider the **azimuth** angles the better. So if we were to take another sight right now, after only these few minutes which it has taken to work up this sight, the two **azimuths** would be so nearly the same that we would have two nearly parallel lines, with such a long overlap where our two **lines of position** came together, that the blur of their intersection would be many **miles** long. We will have to wait at least an hour, better two, to get any kind of good separation. Naturally, the very best intersection would be an **azimuth** 90° from our present one (as is possible with **stars**) but 45° or even 30° will do very nicely. And our small ship will have traveled such a short distance in those several hours (10-15 miles) that when we get our next **line of position,** all that we have to do is **shift,** or **advance,** this first **line of position** along our **course line** that we have traveled, exactly the number of **miles** which the **taff log** will read between the two sights. And this **intersection,** where the second, new **position line,** and the **advanced,** former **position line** cross, will then be just **where we are.** And this, because of shifting that **first position line,** we call a **running fix.**

And that is the beauty and tremendous advantage of taking a fix from a combination of the **Stars,** the **Planets,** and the **moon.** Then you can shoot **two,** or **any number** of celestial bodies at the **same time,** in a matter of minutes, so that your ship moves hardly at all (not enough to consider). And by plotting **all** these **position lines** simultaneously on the chart at one and the same time, man, that is really where you are! And this you can do twice a day, at dawn and evening, whenever you can see a few **stars,** and the **moon** — or even the **setting sun** crossed a few minutes later on the **north star,** though the planet **Venus** is as good, or better than the **Sun.** And there are even those rare moments in the daylight, when you discover the **Moon** in mid-afternoon, like half an old shell hiding in blue sand, and then, if you are lucky, you can shoot it and the **Sun** together and get a **true fix.** But, all in all, you will find the old **Sun,** pretty much the wheelhorse of your navigation. In clouds, in heavy seas, even in rain, it is usually there, easy to find, big, strong and reliable.

CHAPTER II

A Bit More on Time and the Almanac

TIME IS A SUBJECT that most books on navigation make an enormous to-do about. They stir it into one of the greatest points of confusion and discouragement to the eager beginner. There is very little about it that you need let bother you. You will hear and read a great deal about different kinds of **time,** Local Time, Sun Time, Ship Time, Apparent Time, Mean Time, Clock Time, Sidereal Time, the Equation of Time, Civil Time, Nautical Time, to say nothing of our old Greenwich Time, and **plus** or **minus** Zones of Time — but never any mention of **time being just time.** Enough to make you dizzy. And how they love to pour it on!

Time is Time, and overboard with all the Confusionists!

However, there are a few aspects of Time that we should consider, and of these a few are worth remembering.

One thing that you are going to hear a lot about is the **"Conversion of Arc to Time,"** and vice versa. There is a fine 360° Table of this, on the very first of the **yellow pages** in the Nautical Almanac. It is always there, if you ever need it. And what is meant by this impressive phenomenon? Exactly this. You ask me, "How long is the drive to town?" I answer, "About ten **miles.**" Or I can answer, "About fifteen **minutes.**" Both answers tell you exactly what you want to know — how **far** it is into town!

Which simply means that you can always express **distance** in relation to **motion,** in one of those two ways. Our planet Earth turns a certain number of degrees (360°), or **nautical miles** at the Equator (21,600) in a certain number of hours (24). Which breaks down to exactly 900 mph. A fairly good wind-up. Naturally, it is we that are doing the turning; but it becomes easier (granted a bad habit) to say that it is all the other celestial bodies that keep whizzing by at that speed. All that is important is the uniform and continual relation of those **degrees and minutes of arc,** to **hours and minutes and seconds of time.** Which means that in each hour we turn, or the sun and stars appear to move by, 15° of Arc. Or breaking that down still further, one **degree of arc** is traveled in every **four minutes of time.**

How, and where is this useful?

One example. We drift Westward around the world away from Greenwich, where the Sun is overhead at their Clock (Mean Time) Noon. If we kept our ship's clock always set at Greenwich Time, when we were about sighting the American Coast, the **sun** would be **rising** at Noon. Thus, to keep the **Sun** fairly well **overhead** at Noon, we have to keep setting our ship's clock back. How much? Can you tell me? I've already told you. Right. **One hour** back for every **fifteen degrees** of Longitude that we travel West (One Hour forward in East Longitude). So then, if we sail with a fair Trade Wind roughly two degrees of Longitude (120 Miles) in every 24 Hours, how much later must we figure the **Sun** will be **overhead** today than it was yesterday, for our Noon Sight, if we take one? Think now. Right. **Two** degrees, **four** minutes each degree — so, **eight minutes** later in **time** every noon.

And let me tell you, that's about all you ever have to worry about the matter of Time. A little watch setting. A moment of extra calculating, once in a while, for so many additional **miles** of East or West Longitude beyond some Full Hour Meridian.

Full Hour Meridian?

Full Hour Meridian is any Longitude Meridian that is a multiple of 15° **East** or **West** of Greenwich (15°, 30°, 45°, 120°, 330°). **Meridians** of a **full hour** of time change. Which is precisely this business of Conversion of Time to Arc, that we have been talking about. The Nautical Almanac would weigh about sixty pounds if all its valuable data were tabulated for each **degree,** or every **four minutes,** for wherever you happened to be in the world. They just couldn't do it. So, just about all the information that **is** in there, is for the **Prime** Meridian of Greenwich, or (and which is the same thing) for **Full Hour** Meridians in **local time.** If you are somewhere between two Full Hour Meridians, and you know how many **degrees** East or West you are from the nearest, you just have to monkey it out for yourself. Enough on Time for the moment. We'll have a little more later.

TRIGONOMETRIC EXPLANATION
for the Mathematically Curious Only
DEMONSTRATING THE SIMPLE TRIGONOMETRY WHICH THE 249 TABLES HAVE SOLVED FOR US

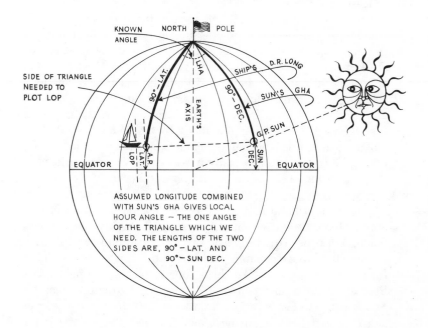

You remember, somewhere around the sixth grade, you learned that, given two sides and any one angle of a triangle, you could find the other two angles and third side. That is all you are doing in solving a Line of Position — only very painlessly, as the tables do all the real work for you. Before the tables, each navigator did this work himself.

We are now going to explore a bit further that double **white** page in the Almanac, which in layout is a duplicate of every other **white** page in this invaluable book. (Incidentally, if either you or your wife is prone to "throwing fits," better have two Almanacs on board; though as a last resort you can always figure the GHA of any celestial body from Table 4, in your 249 Book — at least until the year 2000.)

Except for the right hand half of each right hand page, all the columns give information exactly similar to that we have used for the Sun, only it is for the **Moon**, **Stars** and **Planets**, which we shall presently take up separately. You will notice that the **moon** appears a little more complicated than the others. No wonder — spinning around us, while we spin around the Sun; and now we are trying to render it even more whacky by shooting at it. (You'll be amazed at night, out at sea, by the number of satellites, looking just like bright **Stars** or **Planets**, zooming about, messing up all the constellations. Really makes you scratch your head. If they didn't move so fast in erratic lines, they really would louse things up.)

On the right hand half of the right hand page is a lot of data about Moonrise and Moonset, that I don't think you'll use very much, unless perhaps you have a whiskey still on board. But just to the left of those columns, and below them, is some very helpful information. Let's take those four columns relating to **Twilight** and **Sunrise** and **Sunset**. It's always sad to know when the Sun is going to leave you each day, and nice to know when you can expect it back again next morning, but that's not what is really valuable here.

Those other two columns called **Twilight,** the upper ones for **Dawn** Twilight, the lower ones for **Dusk** Twilight, are really important. They tell you twice each day when to be up there and ready to shoot off a round of stars to really find out and pinpoint exactly where you are. You will notice that the left hand, dark-figured column, under the heading **Lat.**, goes both ways, North and South, starting from 0°, or the Equator. Depending on your ship's exact Latitude, you may have to interpolate this slightly. Or, if you don't want to do that, just get up there a few minutes ahead of time, for essentially what you are after in preparation for any evening or dawn Starsight, is to be ready and waiting for that first moment when you can see the brightest stars and still see the clearly defined horizon, to start shooting and keep going, until the horizon no longer is sufficiently visible to be accurate.

Referring to the above total period of working time as **Twilight**, then **civil** in the columns means the **beginning** of that period; **nautical** means the **end** of that working period. The Time given is Greenwich Time. So, if you keep your ship's clock set each day to Sun Time, advancing it at Noon, that is your **Local Mean Time.** For predicting these Twilights it is the same as Greenwich Time. Otherwise, you will have to keep interpolating your Longitude from the last Full Hour Meridian, which is the method I use.

One last word about Time, and we'll set it back on the galley stove and let it stew for a while.

On the right hand page, directly under these **Twilight** columns, you will find a little block called **Sun.** This is a great little spot. Primarily, it gives you the Time of **Meridian Passage** of the **Sun** for those three days, which means when it is highest, or exactly overhead, or precisely what we call **Noon.** You can see that this is of great importance, if you intend to take a Noon, or Latitude Sight, as this sight is only meaningful and accurate if taken exactly when the Sun is highest, at its daily Zenith. Here again the Time given is G.M.T. To arrive at your Local Time, very careful interpolation of your present Longitude in relation to the nearest Prime Meridian is necessary. Moving Westward, each degree that you have passed the last Full Hour Meridian, will cause Meridian Passage to come exactly **four minutes** later; each mile of the **last degree, four seconds** later. That is the easy way to do it. I defy anybody to do it with Zone Times. Stay away from Zone Times.

And now, I know I'm going to have to tell you what that **Equation of Time** means, in that same little block, just to the left of Meridian Passage. Otherwise, you'll think I'm holding back on something. So here we go again, stirring up that old pot called Time.

What actually makes Time is the Sun. Our traveling around it once every 365 days. And those intervals of light and darkness as we spin on our own axis every 24 hours, in making this long trip. But because we make a huge **ellipse,** not a circle, around the Sun — like a racing car slowing down at the banked turns and speeding up in the straightaways, our elapsed time for each day's spurt varies. Or the interval from Noon to Noon in Sun, or Apparent Time, varies, in relation to our distance from the Sun. If we all lived and operated by sundials, there would be no problem. These lags and speed-ups would take care of themselves and everything would equal out at the end of the year.

But you can't make rubber clocks, with lags and speed-ups in them. Or if you could, they would be very, very expensive and never all quite agree. So our clocks the world over keep an even, unfluctuating Time, which in a year equals the total of all the Sun's uneven Time (with the help of leap year). This **unfluctuating Clock Time** is known as **Mean Time** (or **average** Time). If Clock Time could be represented by a **second** Sun in the sky, the two Suns would always be close to each other, usually overlapping, though rarely concentric. Thus, **Mean Time** is **Clock Time**, nothing more.

Therefore, all that the little **Equation of Time,** in those small blocks at the foot of each of those pages signifies, is the **difference**, given for each Midnight and Noon, between **Apparent Sun Time** and Greenwich **Mean Time,** which naturally applies to any Full Hour Meridian for that day.

So now, are you more confused, or less?

I hope less.

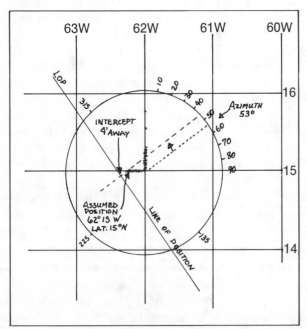

Plotting your line of position

CHAPTER III

Latitude by the Noon Sight

FROM AN UNKNOWN TIME long before the voyages of Columbus, on through the Great Age of Exploration of our then frighteningly vast and mysterious planet, until about a hundred and fifty years ago, ships had no way of determining Longitude at sea, other than dead reckoning. And so for centuries the only method of extended navigation was to **run down a Latitude.**

Let us suppose a ship was sailing from England to a particular island in the West Indies. It was far too dangerous to take the shortest route, to try to sail the direct **Rhumb Line** (any straight Course Line from one place to another on the Mercator Chart). For with the variations of wind and current and leeway, a ship in that distance could be miles off its presumed dead reckoning course on arrival. The only safe practice was, as soon as possible after leaving the port of departure, to sail in safe open water far South, each day at noon checking the Ship's Latitude, which then could be done absolutely accurately with nothing but a sextant, until the exact Latitude of that destination island in the West Indies was reached. Then, each day holding that Latitude by continuous daily Noon Sights, they would **run down that Latitude toward the West,** watching for every slightest sign of approaching land, birds, seaweed, shortening sail at night, if there were no moon, as the probability of land approached — but above all keeping a lookout aloft — until suddenly the cry rang down from the man on the Royal yardarm, "Land ho!"

As early as 1530 a Flemish astronomer named Gemma Frisius, in a work published by him on navigation, first expounded the possibility of determining Longitude at sea by the use of a clock.

(Our old Conversion of Arc into Time). But the only clocks known to the world then were cumbersome pendulum clocks, run by descending weights. And you can well imagine what would have happened to those swinging weights and violently banging pendulums on a rolling ship at sea. Somewhat like trying to play billiards in the howdah on the back of a swaying elephant.

Despite the hopelessness, a century and a half later a Dutchman named Huyghens even made and tried several such pendulum clocks, testing them at sea. After a day or two out of sight of land I doubt he could have found his own backside using both hands. He might have saved himself a lot of trouble by trying his clock out first on the arm of a windmill.

Not until the English, in desperation, offered a prize of twenty thousand pounds ($100,000) in 1714, was the problem of determining a ship's Longitude at sea finally cracked. This meant constructing a clock that would keep Time within three seconds a day on a pitching, heeled-over windjammer, when, at that time, the best pendulum clocks ashore couldn't come anywhere near such accuracy — a fairly safe bet for the British Crown. Yet an Englishman built one! And it doesn't show that vaunted British "fair play" in a very favorable light to add that the self-taught Yorkshire carpenter, who invented and constructed such a clock and won the prize, had to wait three decades, until he was an old man about to be "gathered," before the government finally broke down and paid him his just reward — thirty years after their ships-of-the-line and Indiamen had been using his clock. Yet, until late into the Nineteenth Century, thousands and thousands of ships were still **running down Latitude** by the **Noon Sight.**

Actually, the Noon Sight of Meridian Altitude is about as simple a process as could be imagined. All the information that is necessary is a reasonably accurate knowledge of the Sun's Declination at the time of the sight. If the Sun had no Declination, which is another way of saying, if our drunk little Earth spun straight up and down on its axis instead of listing over to one side (for half the year receiving a bad sunburn on its abdomen, and during the next six months blistering its brisket, which is all we ever mean by Declination), there would have been absolutely nothing to confuse those mutinous crews at all. In all probability the entire business of going to sea would have ended in an abrupt blood-bath, so we can be thankful for Declination.

If the Sun constantly arched each day directly over the Equator, all that these early navigators anywhere North or South of the Line would have had to do, when the Sun was at its high point at Noon — the only accurate moment of measure without relation to any time — was find their angular distance North or South from it to know their exact Latitude, or distance from the Equator. So the one complication that Declination throws into the problem is, that we are never measuring the Sun directly above the Equator except for two days each year — March 21st and September 23rd. As you know, these moments are called the Equinox, or the one time each year when the Sun passes from North to South Latitude and then returns. So the added chore is simply to relate the Sun's Position for any Noon Sight back to the Equator. Which just means adding or subtracting, as the case requires (see diagram page 31) whatever the Sun's Declination for that day is.

If the above isn't clear, get a globe. Let the supporting plastic circumference frame of this globe represent the Sun's daily passage. Adjust the globe so the Equator all the way around is in line with this plastic frame. This is the Sun's path with **no Declination** — one day in the third week of March and September. Now, if you are somewhere North or South of the Equator, you would just measure from that plastic edge and presto — you have your Latitude.

But now tilt the globe, toward or away from you, until the plastic edge, which is still the Sun's direct line of travel, rests about 23° North or South of the Equator. With your ship still in the same place it was before, you still can only measure to that plastic edge, the Sun. You have no other point of reference. But you still want to know your distance from the Equator. But suppose someone tells you, for every day in the year, the exact distance the Equator is from that plastic rim. Sure — you measure to the plastic rim (the Sun with your sextant) and you take the distance your Almanac gives you from the plastic to the Equator and, depending on which side of the Equator you are, you add or subtract that distance and, presto — Latitude.

So, what do we actually do at sea?

Open the Almanac to the proper date. Note the time of Meridian Passage and get up in your shooting-box a few minutes ahead of time, interpolating by adding or substracting four minutes for each degree of Longitude (and four seconds for each additional mile, if you want the exact instant) that you are West or East of the nearest Full Hour Meridian, to which your watch has been set, to determine the precise moment of Meridian Passage for your ship's Meridian, or Longitude. Then, from the Sun column of the same page, starting at 12, or Noon, hour of GMT again move up or down in the column **one hour** for each 15° you are West or East of Greenwich (interpolating if you want to be scrupulously accurate the fractional part of an hour for your minutes of Longitude), and write down the Sun's Declination, naming it North or South. That is all you need from the Almanac.

A few minutes before Meridian Passage, get the Sun in your Sextant and keep following it **up,** by very slightly adjusting the micrometer knob to keep taking up any gap between the Sun and the horizon. The moment it seems to balance, to hang there, watch it closely without touching your sextant again (and you should hear your First Mate shouting "Mark" at this moment if he is checking you with the exact Time of Noon) until you are sure the Sun has started to go **down,** the other way. Read and correct the sextant angle (Hs) for both **Main** Correction and **Dip.**

Now, if the unknown, obviously confused, old gentleman, who laid out the first charts, had set up Latitude with Zero at the North Pole, instead of at the Equator, you would be all set to combine your sextant angle (Ho, when corrected) with the Sun's Declination, to arrive at your Latitude and plot it on your ocean chart, with the horizontal line bisecting your Course Line. You would have your Noon Fix and quickly open a can of warm (bilge-cool) beer. But Latitude, on the chart, happens to be laid out upside down for the Sun. When you are plunk on the Equator, with the Sun approximately directly overhead, your sextant reads 90°, when it should read 0° to be compatible with the chart. (But this slight headache at Noon pays off in the evening because, for the **Polaris,** or **North Star,** Sight, also for Latitude, which you can take twice each day North of the Line, everything is perfect. Your sextant and the chart are right in the same groove.) But, for the Noon Sight for Latitude from the Sun, what do we do with this sextant angle which is just backwards?

We turn it inside out. We **subtract** it from 90°, and call the remainder **Zenith Height.** And use the **remainder,** which then fits right in with the chart scale of **Declination** and **Latitude.** To continue our sight then, **subtract** your corrected sextant angle (Ho) from 90° for **Zenith Height** and **adding** or **subtracting** the Sun's **Declination,** as I have explained, the result is your exact **Latitude.**

Now remember this!

If your **Declination** and **Latitude** have the **same** name (North or South) you **add** the **Declination** to your Zenith Height (or Dis-

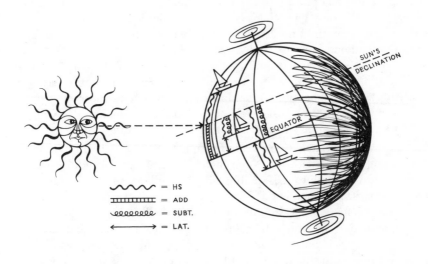

LATITUDE BY MERIDIAN ALTITUDE

tance) and give the **Latitude** the **same** name. If their **names differ,** you **subtract** the **Declination** from the Zenith Height and **name** the **Latitude opposite** to the **Declination.**

There is only one other combination. Your ship lies somewhere North or South of the Equator, at a **lower Latitude** than the **Sun's Declination.** You are between the Sun's Declination and the Equator which you would surely know, because the Sun then would bear North while your ship was still in North Latitude. In this seldom encountered combination, you merely **subtract** your Zenith Height from the **Declination,** and the **remainder** is your **Latitude.** So, that's it, that's all there is. Let's Mutiny!

And now that we've sweated through all this, and whipped it, and come out with the Noon Sight by Meridian Altitude in our teeth, I'm going to risk spoiling all your enthusiasm by commenting on it for a moment.

The Noon Sight by Meridian Altitude has its merits. It is quick. It is accurate. It is fine to teach to an interested crew (preferably female) who wants to doodle with the spare sextant. But it is old-fashioned. It is outmoded in that it has to be done at a certain instant each day; if you miss it, you are through until the next day. One cloud in the wrong place, and you've had it. Also, you have to sit up there, too often in wet weather, waiting for the Sun to hang for as long as ten minutes. All of which is to say that the same Noon Sight, for the same exact Latitude, can be taken at any time around Noon, simply by shooting and timing the Sun and using the 249 Book for a Line of Position.

You come up with exactly the same Latitude, with these advantages. You're only up for an instant, anytime just before or after Noon. So in regard to the time of working up the sight you're probably ahead. You minimize the risk of getting your sextant wet. If the Sun just peeks out for an instant around Noon, you've got it made. And, if for any reason, you don't like the first sight, shoot it and time it again. Or for extreme accuracy, shoot two or three sights, one right after the other, and average them — which a lot of split-hair navigators on small ships prefer to do for every sight, because of the constant difficulty of an accurate horizon. This due to the motion and the lowness of the yacht in the water.

So much for our Noon Latitude by Meridian Passage.

A tot of grog to all hands!

CHAPTER IV

To Shoot the Planets, the Moon, the Stars

THE GREAT PLEASURE OF NAVIGATION is the varied and won-
der-filled half of it that you do at night. It is also during this
dark, brilliantly shifting period, forever changing like a vast moving carpet scattered with intricate patterns of golden nuggets,
that your **True Fix**, composed of Position Lines taken from a
number of celestial bodies almost simultaneously, is possible.
These bright points of reference are all your friends, shining,
winking constellations, arching over and comforting you night
after night. There is nothing about their numbers or their exciting variety that should not inspire you.

Let us begin with the **four** big bright **Planets,** literally street
lamps, for unlike the Stars, they derive their reflected light from
our Sun, as do we and our Moon. In size, as seen in the night,
and movement and luster they are mavericks, halfway in bril-
liance between the Stars and our Moon. We shall begin with
them, for, important as they are, we can despatch them in about
one paragraph.

On the left hand page of your daily **white** pages in the Nautical Almanac, you will find **four** columns of GHA and Dec. (identical to the Sun column in function) for the Planets, **Venus, Mars,
Jupiter, Saturn.** Except for **Venus,** which is always quite closely
anchored to the Sun, so that when you see it, it is briefly each
day in the area just where the Sun will rise, or has just set, the
other three roam aimlessly about across the heavens, traveling
from East to West. **Venus** is the brightest. Then in order of magnitude come **Jupiter, Saturn** and **Mars,** the last distinguished by
its unmistakable reddish color. An easy way to identify any one

of the last three, is to relate its position on any particular night to the navigation Star nearest to it, whose name you will know. Then look up in your Almanac the GHA for that Star for that hour and see which of these three Planets has a GHA roughly similar for the same hour.

To shoot any of the Planets, and plot a 249 Line of Position from any one of them, you exactly follow the procedure for the Sun, on the same Work Sheet blank, with **two** exceptions.

First, and obvious, your **Main** Correction of your **Apparent** Sextant Angle, or Hs, is taken from the **second** (No. 2) or **middle** column, inside the front cover of the Almanac; and as you will notice, it is always **minus** (contrary to the Sun's plus correction). Thus, when you combine this **minus Main** Correction with the **minus Dip** correction, you always get a **total minus** correction, to be subtracted from your Hs. The right hand partial column of Additional Corrections, included in the **Stars and Planets** block, I think is self-explanatory.

Second, on the daily **white** pages, at the foot of each of the **Planet** columns, you will notice a "v" correction, as well as a "d" correction. The value for this "v" is found in the **yellow** pages, in the same space, in every way an identical operation to taking out the "d" value, except the "d" value so obtained (plus or minus, as indicated on the daily pages) always is applied to the **Declination** minutes of arc of the Body, whereas the "v" value so obtained (always plus if not indicated minus at the foot of each column) is applied to the GHA total of degrees and minutes for the Planet, requiring therefore **three** entities for the **GHA** of any **Planet,** whereas only **two** (Hours and Minutes) are required for the **GHA** of the **Sun.** That's all.

Now, having knocked off the easiest of these heavenly bodies, let's take the most complicated, just to prove there's not a great deal of difference between the two.

The **corrections** for the **Apparent Altitude (Hs)** of the **Moon,** together with its **Dip Corrections,** take up both sides of the inside of the **back** cover of the Almanac. But please don't let this frighten you! Essentially, the only difference between the Moon and the other bodies is in this **Main** correction. The only trouble with the Moon is, that it jumps around so each month, and is such an erratic old lady, that its corrections, which couldn't be simpler, cover a very wide range, though always handled with the same brief procedure. The Dip Table needs no comment. Use it the same way as the Table for the Sun, only **please** always take it inside the **back** cover, not the front.

Having **first** corrected your Apparent Sextant Altitude (Hs) for **Dip,** find on one page or the other, across the topmost head of the columns, the correct column for your **Dip-corrected Hs** in **Degrees.** Then, crossing this selected column from the outermost left or right hand columns, which are for the **single** degrees and **minutes** of your Hs, find in the **upper half** of the two pages your proper **Full** First Correction, which is always in **minutes,** and write it down. Remember which column it came from.

Now open to the correct daily **white** page, designated by the date and, for the proper GMT in hours of the "Mark" of the sight, write down the GHA, "v," Dec., "d," and HP values, and don't let all these seemingly-confusing little figures panic you. (HP you can call Horse Power, though it actually refers to Horizontal Parallax, which sounds like a type of paralysis, but is really only the distortion caused by light coming from the Sun and striking the Moon, while being seen by you on Earth.) So, with those values written down, for another moment, let's turn to the **back** cover again.

The Lower Half of both **back** cover pages is for the HP correction. In each of these HP columns, you will see there is a value given for **L** and **U.** This means **(L) Lower Limb,** or bottom edge of the Moon having been brought into contact with the horizon, at the moment of "Mark," or **(U) Upper Limb,** having been brought into contact with the horizon, in taking your sight. I don't remember ever having used the Upper Limb in a sight; nevertheless, simply because the bottom of the Moon, somewhat like a too-ripe cantaloupe (or is it cheese?) can go rotten, you **can** use the Upper Limb. But if you **do,** you must subtract an **extra 30'** of correction, from the **Total** Corrected Sextant Altitude. (Maybe that's why I've never used it?)

So let's assume it is a Lower Limb sight, as I'm sure most of your observations of the Moon will be. Go to the same column that you just took your Upper **First** Correction from, and drop down into the same column in the "Horse Power" (HP) field. Find the nearest equivalent to the HP value you wrote down from the Daily page, which is also always in **minutes,** and **add** this, together with your upper **First** correction, to your **Dip-corrected** Hs, and you are entitled to a slight sigh of relief. Yes, that is your total Hs correction, making your Moon Sight Altitude now Ho. Ahhh!

The rest of the Moon Sight is worked up exactly as for a Planet, regarding the "v" and "d" corrections, "v" plus for GHA, "d" plus or minus for Dec. (Don't let that S.D. at the bottom of those columns ever worry you. In case you want to know what it means, it relates to Semi-Diameter, so now you can forget it.) Naturally, when you get into the **yellow** pages, take your increments for the **minutes and seconds** of your sight from the correct column, in this case the Moon. It makes the sights come out so much better!

And so, the **Aries** column, on the extreme left of the Daily **white** pages, which I know you have been wondering about, is really all that remains to be clarified (with the exception of a few Stars). But first let me suggest that you try not to shy away from using the Moon, just because of these extra little gimmicks. It is a wonderful old friendly yellow balloon and can be of enormous help to you. Just accept its challenge! Incidentally, under the light of the Full Moon, when the sea is calm and you can see the horizon all night, I take **Star Fixes** any time through the entire darkness — if I want to. You will find experts who say you can't do this, that the shine of the Moon adds a distortion to the horizon. This might conceivably be true, if you only took one sight from one Star. But if you shoot a "wheel" of three or more evenly spaced Stars around the horizon, whatever shine error there might be evens itself out, or equalizes, in the tiny triangle of the intersection of your three or more Lines of Position. So, here we go, to the most exciting of all, the Stars.

There are 57 of these navigational beauties, about half of them visible to you at any one time on a clear night. As long as you have got these friendly Stars, they are just like so many nearby lighthouses telling you where you are. But, to begin with, we're going to discuss two terms: **Celestial Equator** and the **First Point of Aries,** both, as always (at least in this book) quite simple.

The **Celestial Equator** is absolutely nothing more than our **Earth's Equator** (or its plane bisecting our Earth) projected upward and outward into the night. Isn't it stupid that we have to mention something so elementary — that the **Equator** in the **sky** is laid out directly above our **Equator** on the **Earth?** But our Gentlemen of Confusion try to make a great difference between these two so-called different Equators. If there is any difference between them (maybe there is) I'm going to let you figure it out.

The **First Point of Aries,** usually referred to by a little symbol resembling a **Ram's Head,** or just a Ram's two curling horns (♈), serves only one purpose — a point of beginning of Longitude measurement in the Sky exactly as 0° Greenwich is for the Earth, as if we drew a red line through the Stars there — for the **Point of Starting** to measure the **GHA** of **all** the **Stars**; or **Zero Degrees** (0°) when that **Point** in each 24 hours passes over Greenwich. This particular Point was chosen because it has to do with the relation of our Earth and the Sun at the **Vernal Equinox,** a moment of no Declination of the Sun, when the tilt of our Earth is exactly neutral to the Sun. If you can think of a better Point to start 6,000 Stars whirling through 360° in 24 hours, let's hear about it — there's never any reason not to change things for the better in this world.

But as far as your navigation is concerned it is not at all important even to know where this Point is among the Stars. You can find it on any star chart as 0°. And it does happen to cross the heavens perpendicular to the Celestial Equator at a place where four distantly-spaced Stars are more or less on a line. But essentially it is only a reference point for our book work. As soon as we start describing the interrelated positions of the Stars in the night sky, I will tell you just where it is. Until then it is just a reference point.

So this beginning for the measurement of Stars is an arbitrary point in the Almanac known as Aries, from which the GHA of all the other Stars is measured, simply by tacking on for **each Star** its additional distance **West** of this **Aries Point** (remembering that the GHA of any body is the angular measurement in degrees from East to West, swinging from a point at the North Pole, from 0° at Greenwich through 360° back to 0° at Greenwich.

In this way we have just **One Base Table** in the Almanac for **all** the **Stars.** This one Base Table is Aries, to which, for any Star, its angular distance measured westward from this Point, called its SHA (Sidereal Hour Angle) is merely **added.** Whoever thought that one up had a terrific idea. Otherwise, for each Star we would have wound up with a separate column in the Almanac, similar to those for the four Planets. So with 57 more columns on each page, you would have had to use the mizzen halyard to get your Almanac up into the cockpit. If this is all clear, there is very little left to the wonderful sport of shooting Stars.

Just remember, if your **Total GHA** for any Star (the combination of its **SHA** and the **GHA Aries**) ever adds up to more than 360°, as frequently it does, simply subtract 360° from it and go right on, and never think about it again. Incidentally, if I don't load this small book with enough actual Work Sheets, or worked-out examples for you (which I always find, when they're cold, when you haven't worked them up step by step from the beginning, are hard to get into) and you'd like to study a few more specific examples of Sights of any of the celestial bodies, there are a number worked up in both the 249 Books and the Nautical Almanac, in the explanatory texts of each. But best of all is to get out where you have a water horizon and start scratching up your own.

With the slight differences stated above, primarily the additional inclusion of the **SHA** of the **Star**, added to the **Hours and Minutes** of the **GHA of Aries** — from there on out Stars are worked up precisely the same as a Line of Position for the Sun and plotted in the same way. Your ship's Longitude in compatible **minutes** is combined with this **Total GHA** of the **Star** to obtain an LHA (Local Hour Angle) in a **full** degree, which is always what you need as argument to enter the 249 Book. From there on, all is identical, as though for the Sun. The comparison of the corrected Ho with the **Total Hc** (Computed Height) from the Tables. If the Ho value is **greater,** your intercept is **toward** the **Star; away** if the Hc is **greater.** And the plotting of the **Assumed Position** (that Adjusted Longitude to satisfy the **minutes** of GHA), the laying off of the **Azimuth** from it, and the final **Line of Position (LOP)** drawn perpendicular to this **Azimuth,** are all routine.

On each left hand Daily **white** page (the Star Page, so to speak) is a complete vertical list of **All 57 Navigational Stars,** so called because of their brightness and their useful position in the heavens. For each Star in this list the **SHA** (Sidereal Hour Angle, "Sidereal" meaning "Star") and its **Declination,** are given. The Declination for any Star, unlike the Sun's, remains so constant throughout the year, that it remains the same. Hence, no "d" increment is ever necessary for any part of an hour, so that is something to be glad about.

It may take you a few weeks, it may take you a year at sea; you may have to cross the Equator a time or two, but if and when you learn to identify and **use** a third of that list of 57 Stars, my friend, you will have it made. You will have the finest, always available, never-ending combination of reference points, to guide you from any port of the seas to any anchorage, with infallible exactness — the rest being up to you and your ship. You will need no radar, no DF, no loud-hailing bullhorn. You will be free and at large on this planet, to set your course whither you wish to go, to know you can arrive there with supreme confidence.

So, if I have made all this clear to you, and if you are still with me, let's begin to explore the whereabouts and the names of a few of these wonderful Stars and how to bring them down to our plotting table.

CHAPTER V

Latitude by Polaris

WE HAVE JUST BEEN LEARNING about the First Point of Aries, so here is a good chance to use it. From those Scout and camping days of our youth, we are all familiar with what we then called the North Star, and how to find it by extending the line of the Two Pointers of the Big Dipper. But to navigators, those are bad words. Especially astronomers don't like those two names: to them the North Star is always and only Polaris. And the Big Dipper is correctly referred to as Ursa Major, which means the Big Bear. So if you will Big Bear this with me, we will use the name Polaris from now on, just to make ourselves feel superior (even if occasionally we still do mention the Big Dipper).

Polaris, at the moment, happens to be the Star nearest to being directly above the **top dead center of our** rotating, or spinning, **axis.** It just misses being "spot on," as the Australians say, by about sixty miles – which is still pretty close. Polaris hasn't always been in this useful position, and in time some other Star, probably Vega again, will be our Pole Star. I don't mean that you have to rush right out and shoot it now, or it won't be there – you have perhaps eight or nine thousand years, to make a rather leisurely sight. The reason for this shifting of Pole Stars is, that our planet Earth has a wobble to it. It is spinning much as a top, that began its spin perfectly, and then the cat gave it a tap with its paw. It has become set in a kind of cone-shaped wobble. Still it takes our Earth about 26,000 years to complete one of these wobbles. Maybe we are just slowing down. But for the next few years anyway, Polaris will still be our Pole Star. In the time of the Egyptian Pharaohs a Star named Draconis would have pinch-hit as the Pole Star. The only trouble was the Pharaohs probably didn't know they had a Pole Star. Or even Poles in Poland. I imagine they were fairly "poleless," certainly without telephone poles, or barber poles, or Gallup polls, being limited very likely to just fishing poles.

But to get back to **Polaris,** it is that sixty-mile eccentric miss of being "spot-on," not the wobble, that gives us our problem. By which I mean, if we just shot it straight, as it is, it could give us an error in our Latitude of one whole **degree,** which, for good, close-to-the-chest navigation, is too damn much. Of course, there are two points (in every 24 hours) in that eccentricity, when from the observer Polaris is to one side or the other in its orbit of the Pole, on a line parallel to the Equator, through the point above the Pole, when that eccentric degree wouldn't count at all. Thus, just about everything we are going to do in working up this sight, is by way of knowing where Polaris is in its nightly spin and eliminating that eccentric one degree.

To find Polaris in your sextant, besides using the Pointers of the Big Bear (our old Big Dipper), there is another very helpful way of locating it. Simply set your sextant to exactly your ship's **Dead Reckoning Latitude,** point it **True North** by your compass, and you will see a fairly bright little Star floating all by itself a little above or below the actual horizon in the sextant. Fortunately, Polaris has no other confusing Stars too near it. Have your stop watch set up on WWV, ready and going. Bring that little diamond point of Polaris up or down, being careful to hold your sextant vertical (and naturally as with all Stars all the filters on your sextant are folded back) until Polaris is just resting on the top edge of that twilight black line of the horizon, seen lurking in the left hand open side of your sextant eyepiece, and — "Mark." You've got it! Check the Taff Log, and if you're not shooting any other Stars, you're ready to duck below.

First, as always, we correct our sight, taking the **Main** correction from the **Stars and Planets** column, No. 2, and also the **Dip** correction, from inside the **front** cover of the Almanac, taking the lesser, upper value for both, **minus** in each case. Polaris, being alone over the Pole, naturally and uniquely has no SHA or Declination. Turn to the proper daily **white** page and, for the first time, we are going to use that **Aries** column, not quite as we would with any other Star, but we certainly do need it, because the argument for entering the **Polaris Tables** (the last three **white** pages in the Almanac) is our **LHA of Aries.**

Again in Greenwich Time (WWV from the radio plus your stop watch) we find the full hour in the GMT column at the extreme left hand side of the left hand page, adding to this amount of GHA the additional increment for the **minutes and seconds** of the "Mark," taken from the **Aries** column of the proper **yellow** page. From this Total **subtract** (in West Longitude) or **add** (in East Longitude) your DR ship's Longitude, in the case of Polaris not bothering about adjusting the Longitude Minutes, as we do not have to use an Assumed Position in a Latitude Sight. This is your **LHA Aries,** which, for convenience, you can round off to the nearest **full** degree. It is this **LHA Aries** alone, that we need, as argument, to enter the **Polaris Tables** on the last three **white** pages of the Nautical Almanac. From here on it is very simple, and, should you ever forget the procedure, there is an explanation and an example at the bottom of these **white** pages.

All that we are going to do now is come up with an **additional** correction to apply to your already corrected (from the front cover) Ho Sextant Angle of this Star, which, let us say was 30° 39′ N. Along the very top of one of these three pages, find the column that includes your LHA angle in **degrees,** which, let us say, was 124°. Then come on down into that **first** upper column (A_0) to whatever line in that 0° to 10° makes it exact, as for 124° you would be in the first column (120°-129°) of the second, or right hand page, and then come in on the line four, which would give you (in the 1970 Almanac) an A_0 value of 1° 02.1′. Now, **staying in that same vertical column all the way,** drop down into the Lat. or A_1 part, and as you are closest to Lat. 30°, you will write under the above A_0 value an A_1 value of 0.4′. And now come on down the **same column** into the Month, let us say Feb., for an A_2 value of 0.8′, which you write down under the other two, and then **add** them all for a total of 1° 03.3′.

Following that little formula given at the bottom of the pages, **all** the values for A_0, A_1, A_2 are **plus,** but also you must **always** **subtract** 1° (one full degree). So we merely knock off the 1° of our Total above, which still leaves a plus 03.3′, which we now **add** to our previously corrected sextant Ho, of 30° 39′, to give us a **Totally Corrected** (for that off-center eccentric swing) **Polaris Angle,** which is our exact, and I mean **exact** Latitude of 30° 42.3′. Or for all intents and purposes, 30° and 42 nautical miles North of the Equator. (Which, incidentally, is too damn cold and variable and I suggest we quickly head South.)

I have always found Polaris, due to the refinement of bringing a Star to the horizon, a much more precise and accurate indication of **exact** Latitude than the Noon Sight by Meridian Altitude of the Sun.

And that is all there is to taking your Latitude by Polaris, a wonderful Star, used alone, or with other Stars. Naturally, you can use it only in the Northern Hemisphere. Even when you are a degree or two North of the Equator, it is hard to find, if there are any waves at all. For, understandably, when you are **Dead On** the Equator, it should register 0° on your sextant, which means a very wet Star. But as you come North, it is a great friend — that twice each day, at dawn and dusk, will tell you just what your Latitude is. And what a fine Fix in the dawn, to bag Polaris, then, like shooting two ducks — one with each barrel — swing around and bust Venus, just where the sky is lightening in the East, heralding the sunrise. And there, bang, bang, you have the perfect Longitude Line to cross on your Polaris. Could anyone ask for a finer Fix, with which to start the day?

Hurrah for Polaris!

CHAPTER VI

Identifying the Big Four Constellation Groups and a Number of the Most Important Stars

WHEN YOU FIRST LOOK UP and think about starting to use the Stars to navigate by, you realize that there are a lot of them. Also, they are rather widely spread and look damn confusing. It will amaze you how quickly all this will fall into several relatively simple, broad patterns. And how easily the **Thirty-odd Navigational Stars**, that you will use to navigate by, around the benign midriff of the world, will fit into a general and always recognizable framework. First, a couple of hints.

In the very beginning, in trying to orient yourself alone at sea, you will tend to look for the constellations and arrangements of Stars on too small a scale. They are usually huge, with a vast sweep. Until you suddenly identify, one after the other, these mighty designs, don't try to hunt for elephants with a squirrel gun. Also, begin with just that part which will appear above you in the first hour each night, depending on what month you head out to sea. Become familiar with just those few stars, as these same ones will come out each twilight night after night. Then, as the nights wear on, widen your band, until gradually you become familiar with the full sweep of that season's stars. This will take you an exciting week anyway, and you will enjoy every exploring minute of it.

We have already become familiar with Polaris. And this is a good place to start to work out from. This northern, or Polar area, is small and tight, and has a number of easily recognizable starting points, visible from anywhere in our Equatorial Band at just about any season of the year. Of these, the Big Bear is probably the most well known and easiest to start from. Mind you now, we are only going to try to establish certain patterns in the Stars at first. Some of these later may prove not to be Navigaional Stars, for us, because of their high Declinations (in that they lie too far North or South of the Celestial Equator to be included in our Tables). Fortunately, it just so happens that the greatest concentration of bright easy-to-find Stars lies not too far North or South of the Celestial Equator, and so naturally these are the ones the Tables are designed to use.

A word now about the **Big Bear,** and this **Northern group of Stars** that revolve around **Polaris** in close association with it (for really all the Stars turn around Polaris). Moving from the Bowl of the Dipper toward Polaris, and holding that **Line** until just about the same distance on the other side of the Pole Star, for it lies exactly 180° around from the Big Bear, is a very easy to find group of Stars resembling an elongated and slightly irregular "W." The Greeks thought it looked like a chair. This small bright little constellation is called **Cassiopeia** and is important. Its five Stars are always quickly found and a great help in orientating yourself. And now, to keep a promise, that **Line,** which we just traced across from the Big Bear, after it leaves Polaris and starts on down toward the Celestial Equator, just nicking the brighter, bigger-half edge of Cassiopeia, is our **First Point of Aries.** This Line of Aries continues on down southward, until it hits head-on the next bright Star, which you will discover is the NE corner of an almost perfect **Square,** of **Four** bright **Stars.** In case you have any trouble finding it, this Square is just about the right size to put Cassiopeia inside it, if you were going to frame our "Flying W," better known as the Chair. That Square is the **Square of Pegasus.** The ancient Greeks thought it looked like a horse (maybe Barney Google's Horse, "Sparkplug"?). More likely the way a horse might look to a Greek seen through the bottom of a just-emptied wine glass. We'll come back to this Barney Google Horse later. But for the record, it's the only arrangement resembling a big true **Square of Four Stars** anywhere in the night sky.

To quickly give you the name of your first navigation Star, that NE corner Star of the Square, lying right on the Line of Aries, which you hit coming down from the Pole just touching Cassiopeia, is named **Alpheratz.** If you will bear with me, I have found that the best way to remember the crazy names of these Stars, is by some little catchy "gag" or "pun," and the more out-on-a-limb the better. Such as, if you can imagine this Square made of Four Musketeers (instead of the usual three), friends of Barney's, this

NE one's name obviously would be "Rats." Because then they could say, "All for one, All-fer-Ratz." So, there is Alpheratz — one of the Four Musketeers forming the Square on the Line of Aries. And when he gets tired of holding up the head of that horse, "Sparkplug," "Rats" has only to make a quick little run due North to sit down and rest in the Chair of Cassiopeia. After he's rested, if he wants a cool drink, because he has to cross the Pole to get it, he can find it straight across 180°, waiting for him in the Big Dipper (but that's cheating).

What else about this Polaris, or Northern group, of Stars? There is another Bear, who really looks more like a small tin can with a string tied onto it for a tail, with the end of that tail tied right to Polaris. That's the **Little Bear,** and there's mighty "little" he can do for you, not even growl. He's just there, mostly tail. So, excluding the Square (Pegasus wouldn't stand still for that) that is one group of Stars, of our so-called **Big Four Groups** to help find all the others. So let's use it again now, to broaden out way across on the other side of the night (all the Stars being called the **Celestial Sphere**) remote from Pegasus, which for the time being we will leave with "Rats" and forget. (We went down to pat him just because I wanted to clear up that First Point of Aries.)

We're going back to the Big Bear. And I should mention that very often you can't see both the Big Bear and Cassiopeia at the same time; as they are 180° apart, one may be down just under the Pole. But tonight we can see the **Dipper** and we start on that long handle and we get a good run and we run right off that handle, following its curve, into the night. As the end of that handle is curved, we are going to be traveling in an **Arc** — and away on out we go Southward, curving down toward the Equator on that Arc. Until, we can't miss it, about the distance again of the whole Dipper, we run smack into a big bright Star, with nothing nearly so bright around it. And who is it we find on this Arc? **Arcturus.**

There he is. One of the real work horses of the navigation Stars. **Arcturus.** But don't let's stop yet. Let's keep right on going on that same great **Arc,** curving on just about the same distance that we have come. Again, you can't miss this one — a bright Star **Spiked** almost on the end of the **Gaff** of a little sail shaped like a **Spanker,** which is actually what those Four Stars way down there below the Celestial Equator are called, the **Spanker.** What would you guess the name of that bright Star is, **Spiked** on that **Gaff?**

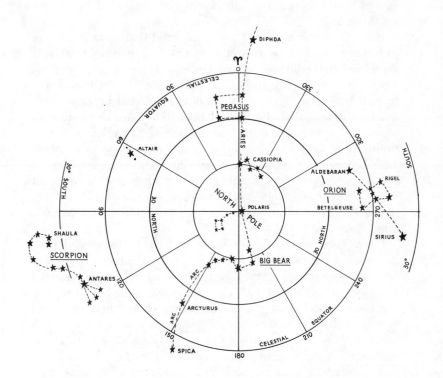

THE BIG FOUR CONSTELLATIONS
KEY TO ALL OTHER STARS

Spica. Right. And that's another good Star to shoot at with your Prayer Wheel. Spica. The only Star Spiked on a Gaff.

Oh, man!

But let's get on with it. Having followed that great Arc we've just made, down through Arcturus to Spica, we're right down in the **Heart** of the 30°-30° of **Navigational Stars.** Because the 249 Vol. II has only space to handle Stars whose Declination is not **more** than 30° **North or** 30° **South of the Celestial Equator,** these are sometimes called **Equatorial Stars.** They are the only ones we are going to talk about in this book. So let's explore out now, in that band, on both sides of that great Arc we've just described. Let's broaden our knowledge to take in a few more Stars, and a few more wet puns to remember them by. Of course, if you want to invent your own puns, you'll remember them a lot better. Or, if you can think of any other way to make them stick in your mind, have at it.

From Spica, we will move forward (East) in the direction an arrow would shoot, if that Arc were a bow, slanting South away from the Equator (for Spica already is well below the Celestial Equator with a Dec. of S 11°) and we come to the **Second** of our **Big Four** constellation groups, the **Scorpion.** And here those old Greeks really did make sense. Perhaps they sobered up for a moment, long enough to name this one. Huge as it is, this constellation truly does resemble a Scorpion. Stinger, whiskers, eye, tail — everything is there. Look for a great parade of bright Stars in a sort of an evenly-curved enormous "S," with the roundly brilliant tail trailing off down toward the South Pole. The northern end of this giant Scorpion, the head, contains an unmistakable bright **Red Star,** the **Scorpion's Eye.** And what an Eye for navigating! Out from this Eye (in the direction of Spica) are three white Stars in a small perfect arc — the always easily recognizable "whiskers" or "feelers." But again, that Red Eye of the Scorpion is a terrific Navigation Star, and we've got to know its name to use it; for if you can't look it up in the books by the right name, there is no sense in shooting any Star!

While the Scorpion was crawling up from the South Pole to attack Spica (helpless on that Gaff), it held its head so low it got an Ant in its Eye — the cause of the Eye being Red. And so, obviously, the name of that **Red Eye** is **Antares.** Incidentally, Antares is just about as far South as our 249 gun can shoot, its Declination being S 26°. (We could go another 4° South.)

Again, let's go back to our **Great Arc.** If once more we think of it as a bow shooting an arrow, the arrow resting low on the bow about midway between Spica and Arcturus, and we let that arrow fly way out, parallel to and slightily North of the Equator, passing well North of Antares in the head of the Scorpion, going on beyond the Scorpion about the same distance that Spica lies to the other side of it, just North of the Equator with a Dec. of N 09°, is a fine lonely bright Star, with the Milky Way far out in the night behind it. On **each** side of this bright Star, evenly spaced from it, are **two** less bright Stars, all three in line with the bright Star in the center, somewhat like three candles on an Altar. This is **Altair** (on the Altar), one of the first-rank great navigational Stars. (Also a name you see on too many yachts. After all, some yachts could be called Zubenelgenubi.) And just for the record, way up North of Altair, off in the direction of the Big Bear, but out of our 30°-30° Declination band so that we can't use it in this book, is, I think, the most beautiful of all the Stars, winking white-blue, like a finely-cut blue diamond, **Vega.** You never want to put your last **Blue Diamond** down on the table at **Las Vegas.** No. Never. Too great a loss of beauty in the night sky. More of Vega, the blue diamond, when we get into that special **Vol. I of 249,** which soon we will be coming to.

And now, just to let you know where you are, if we imagined ourselves standing on the Altar of Altair, and we squinted our eyes, peering way on out, further on in this same direction we have come from the Arc, we should see that old Square of Barney Google's "Sparkplug," held out and propped up by our Four Musketeers, one of whom — the farthest and most Northern from us, you remember, is Alpheratz (known, for short, by the other three, just as "Rats"). So just to kind of wind things up in this vast area we have traversed, the corner of Pegasus (Sparkplug) nearest to us, that SW Star, one of the Four Musketeers, is a rascal called **Markab.** He is nowhere near the fighter that "Rats" is (being of a lesser magnitude), but still ready if you need him. Alpheratz, or "Rats" is, naturally, strong for the "corn," whereas this one, obliquely across from him, is "more for the cob," prefers "More Cob." **Markab,** the country boy. (Heah! Well, you ought to remember that one!) Enough? Then on the strength of that being **enough,** we ought to tie in the closest navigation Star to Markab, it being **Enif. Enif,** for **enough,** lying somewhat closer down toward the Equator and back in the direction of the Scorpion, actually not too far North of a line from Alpheratz to Markab extended out to the SW.

I might mention here that there are two pages of fairly good Star Charts toward the very last of the **white** pages in the Nautical Almanac, though I think the two rectangular ones of the Equatorial Stars would relate to the sky better if they were turned upside down. The two similar charts, shown in this book, show the Stars as you see them, when you face North, looking up over your head. You can also obtain for less than twenty dollars a fine Celestial Sphere, similar to a globe of the Earth, but with the heavens, the Constellations and Stars, printed on it instead.

Before leaving this area, let's take in one more Star on our way. Drop due South from **Markab,** as far as you can go in the ballpark and still use its Declination for our book — right down to the South 30° yard line, like kicking one way down the length of the field and, man, he missed it! He **Fumbled Out!** Because way down there all alone, with nothing around it, down on the 30 yard line, is **Fomalhaut.** He has never caught the ball yet, always **Fumbles Out.** Old **Fomalhaut!**

So, there we have another one. Again, for a little more mopping up, angle back up across the Equator to **Altair,** that middle candle of the three on the Altar. Strike a line from Altair straight across (Westward) back to Arcturus, on that Arc of our beginning. About halfway of this straight line, and still on it, you pass right under the constellation of Hercules. We aren't interested in him, but being that near to so much muscle, what would be more likely than to find a man **wrassling** with a **hog,** particularly when his own constellation is named **Ophiuchus** (I'm guessing that those Stars must have looked like a sneeze to the Greeks.) So there is **Rasalhague,** midway on a line between Altair and Arcturus, below an imaginary Hercules. He can be a good fighter for you, because he is the only Star in a big area there, when you need help through your sextant.

Perhaps we ought to take a breather here, before crossing to the other side of our Great Arc, to chase down a few more of these valiant battlers of the night, and their patterns for giving you Position Lines. To take time out, I'll tell you a really shameful mathematical pun. There are three Indian Ladies, sitting respectively on a Buffalo skin, a Hippopotamus skin, and a Walrus skin. Which one would be the biggest and fattest? Simple, if you just remember that, "the Squaw of the Hippopotamus is always equal to the Squaws of the two other hides." Now you should welcome going back to our Stars.

If we consider the **Square of Pegasus** as the **third** of our **Big**

Four Groups, then we have only one more — probably the most impressive of all. But on our way to it, we want to pick up two or three more scattered warriors, to do battle for you when you need them. Starting from Arcturus (Dec. N 19°) let's cross the Arc and move on out now away from it in the opposite direction (West). Rather quickly we will encounter a very sharp spearhead of three aggressive little Stars, coming toward us like the next arrow to be fitted to the bow of the Arc. The very tip of this arrow is the brightest Star of the three. We ask the arrow where it is going. It shakes its "pointed" head and mutters, **"Den' know** — jus' the **Bow there."** Hence **Denebola!** And as we pass, he goes on muttering — **"Den' know** — jus' the **bow there."** The puns seem to be getting worse, don't they? Anyway, that's its name, **Denebola.**

And right on West of Denebola, in fact all in the same constellation of **Leo, the Lion** (it actually has a tail) is a perfect and unmistakable **Sickle** in the sky, with a fine bright navigation Star marking the **outer** end of the **sickle** handle. **Sickle** being the symbol of the Communist world, whoever is associated with **sickles** must be a **regular** guy. This Star had to be called **Regulus.** And so we have **Regulus,** the **regular** Communist, known for a good tight **Fix** on the handle of the **sickle.** And just below this first Muscovite, across the Celestial Equator and a little further on, though not quite as bright, is another Communist, who grew up in prison butting his head against stone walls, until his **scalp** got as **hard** as a frozen boot. Actually, he hit it so **hard** he knocked the **sc** off his **scalp.** And so we introduce you to custard-skulled **Alphard,** from the other side of the Trans-Siberian tracks — an iron-dome from the word "swat." No **'Alp** is as **hard** as **Alphard's.**

As we approach this greatest of all the Constellations, which we are heading for, we come first to a protective arc, curved the same way as our original Arc. A protective arc made up of **three pairs** of evenly-spaced **twins,** like outriders, or guards (in pairs for safety) protecting the Great Hunter-God, just beyond them. And in each of the two **lower** sets of **twins** we have a mighty navigation Star, the uppermost third pair of twins being out of our Equatorial Band.

Procyon, the lowermost brightest Star in the lowest pair of twins, at Dec. N 05°, almost on the Equator, is a guy "for" or **pro** just about everything. Obviously a **pro** of the first magnitude, he is Pro a lot of things, including any Scion of the great Orion. (If Orion's Scion were a Lion, he couldn't be more Pro Orion Scion than Procyon.) Oh, man! Procyon, in the Constellation of the

Small Dog, the brightest in the three pairs of twins to Orion. And above Procyon, in the next higher pair of twins, and these are actually called **The Twins,** the brighter Star is **Pollux,** his identical twin **Castor,** being just over the N 30° yard line, out of the game for this book. (Probably all for the good if the coach has given him a dose of Castor Oil.) So Pollux is our only other navigational Star of these lower twins. Maybe to get rid of the smell of any splash from that Castor Oil, he keeps continually washing his Poll in Lux. The coach must have given Castor an awful dose, because Pollux never stops washing his hair. Ah, those puns? I hope they help. Maybe they just make you so mad, you can't help remembering a few of these tongue-twisters.

And now we come to the greatest fighting group of Stars of them all for the anxious navigator. The combination of **Orion** and the **Big Dog.** For who ever heard of a Hunter who didn't have a pooch or two at his heels. This one has two dogs, the one farthest from him represented by a single Star (**Canopus**), being just the dog's eye, but enough for those old Greeks to make a whole canine out of it. (What you call really "putting on the dog.") But Canopus is way too far South for us at the moment, even if it is the second brightest Star in the heavens, almost down over the Antarctic Circle. So we will have to be satisfied with just the other Dog, whose Eye is the very brightest Star in all the sky. But first let's start with the famous Hunter, himself.

Those three fine bright little Stars in a row, like a row of three bullseyes Dead on the Equator are Orion's belt; you can even see his Bowie knife — three faint smaller stars hanging down from it. (Actually, that Bowie knife is another entire galaxy, like our Milky Way galaxy, whirling out there in space billions of miles behind Orion.) North of this belt on the Equator are two great bright Stars marking Orion's shoulders, one of them the renowned Red Star, **Betalgeuse,** which the crews on the old square-riggers used to call "Beetle-juice," undoubtedly because of that red color. The other shoulder Star is **Bellatrix,** which because of such a variety of stars concentrated there, I doubt you will ever use. But there is not another Star more helpful, or more easy to recognize in your sextant, due to its reddish color, than Beetle-juice, this upper shoulder Star of Orion. (I've often wondered if Irish Skippers don't insist his name is O'Brian.)

And across from Betalgeuse, the red shoulder Star, obliquely across, forming the off knee of Orion, across the Equator and on the opposite side of this imposing constellation, is **Rigel,** a most needful Star, even if he is a kneecap, suffering from that old ecclesiastical complaint known as Rigel on the knee, from too much praying. But before we leave Orion, let's go back to his belt and squint along those three belt Stars out in the opposite direction from Procyon. Using the belt as a sighter, we see a lonely, somewhat reddish Star, another of the good bright navigation Stars, **Aldebaran.** The **Old Baron.** Alde meaning Old, as in "Auld Lang Syne." Aldebaran, the Old Baron with a red nose. And way out beyond him (really more in line with the pointers of that belt) are a couple of more old gents, **Menkar** and **Hamal.** They must certainly always be hungry and thirsty, because, when they're so far out and alone, always **men care** for **ham** and **ale.** Menkar and Hamal, alone and hungry, out where men care only for ham and ale. And that, believe it or not, is about it — except for the brightest Star of them all.

Serious **Sirius,** the **Dog Star,** bright as the light on a locomotive. A Star with a fire on it twice as hot as our Sun, and of matter so dense it would put gold in the same weight class as a feather. Just use that belt again as a pointer, only now in the opposite direction from the Old Baron, and there is Sirius, the Eye in the Great Dog. When you can't see Sirius, you're in a serious way of being bitten, as far as navigation goes. Or any way you look at it, a dog with an eye that intense and bright, is Sirius. And to **add horror** to the dog, you'd better **add hara** to your thinking, for that's **Adhara,** just South of Sirius, down on the S 29° yard line. (Maybe those imaginative Greeks merely saw Adhara as a mole on the first dog's tail, or maybe it's a tin can tied there, but it surely adds horror to that eye. It really gives that eye a serious look.) But anyway, who would want to aim a sextant at Adhara, when they have the biggest, brightest light of them all growing out of the other end of the same Dog. And, yes — believe it or not — just out there beyond Hamal, that's the other side of "Sparkplug," the Square horse, better known as the Square of Pegasus. And so, we have roughly covered the entire 360° of the ball park. Whew!

So, now all you have to do is lie out there in the balmy darkness of the cockpit, the masts above you gently swaying under the even pull of the warm night Trades with the phosphorous streaming out astern in a great slip-stream of green fire, and just keep gazing up, recognizing them and memorizing their names, one by one.

Oh, what a rough life!

CHAPTER VII

The Use of Vol. I, H.O. Pub. 249

BEFORE I BEGIN TO EXTOL this magic volume in its red binding, let me first say, by way of relieving you of any unnecessary alarm about further effort on your part in mastering the art of navigation, that already you know all you need to know, about the use of the Sun, Planets and Stars to successfully guide your ship anywhere in the world.

What this new book offers is a simplification and time-saving method to augment what now you know. It pertains only to the use of the Stars and their special value in giving you the only True Fix at sea. It is not essential that you ever use this wonderful book (any more than you need a self-starter in your car, if it is equipped with a crank). You can accomplish the same results, the same exactness of Position, by using Vol. II. Then why didn't I explain this book, in the beginning, instead of the other? Because you have to learn the basic groundwork intrinsic in Vol. II to know how to use and appreciate the advantages of this new book.

Volume I does the same job as Volume II, in regard to Stars, merely doing it quicker and with less work, offering more Stars, while at the same time acting as a **Star Finder** to find for you and identify every Star. It is not limited to those Stars only in the Equatorial Band of 30° North and 30° South of the Celestial Equator. So you may suit yourself as to "if" and "when" you may choose to add this valuable tool to your command.

The chief beauty of this book is, that you can do nine-tenths of your work in advance, in preparation for the shooting of a round of Stars before the time of twilight. So that within minutes (actually less than five minutes) after shooting the last Star, you can have them all plotted on the chart and your Fix completed. In regard to time alone, you can have your Fix before you could even have corrected your Apparent Sextant Heights, let alone start the rest of the work, if you were limited only to the use of Vol. II – that great basic 249 Book with the white binding, the only book for the Sun, Planets and Individual Stars. If just this time saving advantage appeals to you, as so often nearing land at dawn, you really want to know where you are fast – hang on! You will be amazed how simple it is to use this new book.

There are many ways and combinations of benefiting by this ingenious method developed in Vol. I, several variations being explained in the introduction of the book, itself. You can also, in conjunction with Vol. II, merely use it as an advance guide, telling you which Stars to shoot, as well as giving you their heights and Azimuths in advance, to help you locate them (actually find them for you). Then you may work up the sights with Vol. II, or, if you wish, work up each Star individually as to LHA Aries after shooting with this new Volume. But I am now going to explain what I consider the best and most helpful way in which this new book can save you time and work.

If you anticipate wanting a Full Star Fix at either Evening Twilight, or the following Dawn – that afternoon when you have plenty of time and good daylight with the pressure off, we are going to get everything ready, even to Plotting our Assumed Position, as well as the Azimuth lines from all our Stars. The moment the shooting is over, we have only to subtract for our Intercepts and draw our Position Lines.

And how do we start?

By extracting from our Nautical Almanac, as accurately as we can, the approximate center, or optimum visibility time, of the approaching Twilight, Evening or Dawn, as the case may be. As you know, this would be the **mid** point between the Nautical and Civil Times given on the correct Daily page, interpolating for your correct Latitude (as indicated by the layout of the page) the correct Local Time from the nearest Full Hour Meridian. Of course, you may already know very nearly the Time of Twilight from the previous evenings or dawns. In any case, what we are going to attempt to do, in using this book to its best advantage, is to set up a Timetable and take our observations at an exact certain timing. In other words, the Timekeeper will shout "Mark," while you wait for him holding each Star in your sextant balanced on the horizon.

Let us assume we are near Latitude 20° North, around the middle of January, make it the 15th to be exact, DR Longitude 152° West (approaching Hawaii).

In the **Twilight** column of our Almanac for that day, we find that **Civil Twilight** for that Latitude begins at 17:38 and that **Nautical Twilight** ends at 18:09. The middle, or optimum, of that Time would be 17:53. But we are 2° further West of the 150° Full Hour Meridian, so we must **add** another eight minutes, bringing us to almost 18:00, dead on the nose, as the middle of Twilight. Our Full Hour Meridian of 150° is ten hours later than Greenwich, right? Therefore, to make sure you follow all this, 18:00 Greenwich, advanced by ten hours, becomes 04:00 of January 16th for our proper GMT in the Almanac of that Evening Twilight out there in the Pacific.

If the above is clear, we will turn to the January 16 page in the 1970 Almanac and take out the GHA of Aries for 04:00, which we find is 175° 10.8′. (If we had increments for Minutes, which we don't we would naturally **add** them to this hourly GHA, taking them from the Aries column on the proper **yellow** page.) From our GHA we **subtract** our DR Longitude of 152° W, as always, making the Minutes and Seconds (10.8′) the same as those of the GHA. Thus, hours ahead of our Sight Time, we now have our Assumed Position for our Sight, 152° 11′ W, that is — if our Shooting Time for One Star takes place that evening exactly at 18:00:00 Local Mean Time. And also, from the above subtraction, we have come up with our LHA Aries for that precise Time, as being 23°, which is all we ever need to know to use our new **Vol. I, red-bound 249 Book!**

You can put the Almanac away. We don't need it for any SHA or Dec. or any Stars, or for any other GHA of Aries or Minutes and Seconds for other Stars in our wheel, for we are going to make the Timing of our other Sights such that we won't need them. In fact, just about all our heavy arithmetic already has been done.

For the first time we open our new Vol. I, to the page headed Lat. 20° N. And there, believe it or not, we find just one column as **Argument — LHA Aires** (indicated by that little Ram's Head symbol). And running down to LHA 22°, look at what we find! Man! **Seven** pre-selected Stars spread right out there for you, each with an exact Hc (Computed Height) and Azimuth, to lay off in advance from our already determined Assumed Position, as well as telling us what Stars we can expect, together with quite a bit more about them!

Those Stars in CAPITAL LETTERS are of the **First Magnitude,** the brightest — which should be taken into consideration if it might be a hazy evening. Those designated with an **Asterisk** (*) are the three Stars suggested as giving you the best angular cuts, in the intersecting of their three Lines of Position. What else does it tell you? Three of those Stars are new to you, as they are each to be found somewhere out beyond the 30°–30° Declination limit of the Equatorial Band, which as you know is the maximum Declination of useable Stars in Vol. II. But you don't know where these new Stars are, so how can you use them? Very simple. Here's how!

The **two** figures given for every **Star** in this book provide you with an **absolutely infallible Star Finder!** Even for those Stars that you already know, it is the quickest and safest way to locate them, to make sure you have the right Star in your sextant eyepiece. Merely set your sextant for each Star at exactly that Hc Height given for whatever LHA Aires your Timing is set up on, and with a glance down at your Compass in the binnacle, aim your sextant in that Azimuth direction and you will see just one bright Star, and one Star only, floating on the horizon right in your eyepiece. These Stars have been selected for just this reason — having relatively **no** other Stars around them to confuse you. Even should you now not wish to continue your Sight by the method of this new Vol. I, instead merely waiting for Twilight and shooting your Stars in the former method, one by one, timing them however the moments fall — consider all the extra and invaluable information you have gained, to this point, in advance. (Of course, you then would not use our pre-determined AP). You not only know the **seven best Stars** you will see, you also know their exact **Height** and **Azimuth,** which, if you are shooting at random, as each four minutes of time passes, you merely move down to the next higher LHA and take your Star Finding figures from it, as I will elaborate on in a minute. But to do the wise move, and continue with the method of this book, here is all you do, to continue on with our Pre-timed Sights.

Let's first take the advice of the book, as given for our LHA 22°, and set up on the three Stars — Capella, Fomalhaut and **Deneb,** as they are each given in Capitals and two with the Asterisk. Deneb is new to you. You will discover it is a fine big

Star, being the Head Star of the great **Northern Cross,** at a Declination of N 45°, its SHA 50°. Anyway, we can right now on our chart Plot from our pre-determined Assumed Position these three Azimuths for these three Stars. The way we are getting ahead of the game, it is certainly not going to take us long, groping around in the dark for the pencil, once we've knocked off our three Hs angles from these three Stars at Twilight.

But, how are we going to use one Assumed Position, for all three of these Stars, or even four or five, if we decide to shoot that many? Our LHA of Aries, which we already have worked up from the Almanac, is for one shot, timed exactly for the Full Hour of 18:00:00. Is there any way that we can use that same Assumed Longitude (within the orderly workings of Outer Space) for three or more Stars, and so be able to Plot them all from one Assumed Position? There is. If you will remember that the Earth travels exactly **one degree every four minutes,** and to prove it, turn in the **yellow** pages of your Almanac to the 4ᵐ Page for Aries — 1° 00.2'. And so with the passage of every additional four minutes of time, our LHA will increase by exactly One Full Degree.

Therefore, if we **timed** our shots, starting with any full minute near the center of our Twilight period, as now we are set up even to a full hour (18:00:00) and took other shots at exactly **every four minutes,** at each one of those intervals we would only be changing our original arithmetic by exactly **one full degree** (of GHA and thus also of LHA). Mind you, in doing the above, we have never altered the minutes or any other part of our Assumed Longitude and consequently have not changed our Assumed Position. It stands to reason then, that we can keep on shooting those Stars, one each four minutes, for as long as we can see them on the horizon, and for each new sight just **drop down one line** in our Tables on that same page (or one line up, if earlier than our controlling Star), where we began, in Vol. I — allowing us, for all these shots, to use the same Assumed Position which we already have plotted on our chart! Is that a saving of time and work? I ask you?

And that's about all there is to it.

Comes Twilight, get up there. Find your first Star in your sextant. Hold it to the horizon until your First Mate hollars "Mark" at exactly 18:00. Read off your angle. Write down your Hs beside the name of that Star. And you've got four full minutes (setting your sextant at the Hc for the next Star, moving down one line in the Book for this next Hc and Azimuth) to find this next Star and bring it to the horizon. When you have shot the three, or more, that you want, **correct each Hs** for **Main** and **Dip,** compare these with the proper Hc on the proper line of LHA for each in the book, advanced for each successive Star by one line, figure your Intercepts (Toward or Away) and Plot your several Lines of Position all from that one Assumed Position! And that's it!

You've got your Fix.

One more last little gimmick that you should take note of, as the years press by, and you drift on and on. At the back of Vol. I are several helpful Tables. Please turn to **Table 5.** As the years change, to be exact and accurate as to just where each Fix really is (for the Earth is varying slightly in relation to these Tables) also using LHA and Lat. as argument, this Table 5 will give you the correction. For any combination of years, etc. the Table will give you two simple figures: the first is the **distance** your Fix is to be shifted in miles; the second is the **direction,** or Azimuth, of this shift, from your arrived at Fix. If you are in a year that requires this slight correction in order to be able to continue to use this book — and again if you want to save time and get your corrected Fix quicker, you can just as easily shift your Assumed Position in advance, as you Plot it, and have done with this shifting beforehand.

Two more suggestions. First. If you use just three Stars in the procedure described in this chapter, it is advisable, as being slightly more accurate, to set up on the **middle Star** in your Timing, advancing the first Star by four minutes, and therefore, for it, moving **up** one line in the LHA column — it being the first Star you will shoot. Then the **middle Star** uses the **original LHA** you have figured, and you move **down** one degree for the third Star. Thus, with five Stars, advance two, let the last two follow the original Timing.

Second suggestion. For those who desire to save still more time in getting that Fix Plotted, at the soonest possible moment after the last sight, you also can do your corrections for Main and Dip for each Star in advance. Substituting your Hc angles, given in the book, for your still unshot Hs angles, as they are always very similar to each other, figure in advance the total correction for each Star, and **reversing** the sign (your correction is **always** a **minus** one for Stars), you make the correction **plus** and **add** it to the "Hc" (instead of the Hs) for each Star. This does not alter your ultimate Intercept in that the amount, or difference of comparison remains the same. All you have to do then, after shooting, is compare each raw uncorrected Hs angle with its reverse-corrected Hc angle in the Tables for Intercept and Plot it. And this, I promise you, is the last rock-bottom refinement for maximum speed in getting that Fix on that chart.

Good hunting!

CHAPTER VIII

Helpful Hints

WHEN ANYONE HAS GUIDED A SMALL BOAT over a considerable amount of the Earth's waters, probably what he remembers most vividly are the near misses. His close brushes with disaster. In short, the lessons he learned from his **own** exposures to danger, and to a somewhat lesser extent, from what he has heard and heeded, concerning the regrettable experiences of others.

No one **ever** will be infallible. And it is true that, at sea, you never come to an end of those **first time** experiences. Every coast and every storm, and every combination of current and tide and bottom is different. And though experience is the greatest preventer, one never seems to have had quite enough of it. Yet certain basic rules and proven advice probably, in this case, are better said than not said, regardless of how questionable the giving of any advice.

I shall set down a list of **Twelve Don'ts** which include the causes of most of the mis-adventures that happen at sea.

1. Don't, under any circumstances, regardless of the urgencies, **close with the land at night.** Even to seeking and entering shelter which is well known to you. By the same reasoning, don't, if possible, time your voyage to pass obvious dangers — islands, reefs, shoals, at **night.** Heave-to. Turn about!

2. Don't lay your course closer than you absolutely have to, in passing dangers — islands, rocks, reefs, — especially at night. If the dangers are lighted, lay your course only close enough to pick up the outer limit of that light. Only to know you have passed it. Ten miles off is a good, safe minimum limit. Nine out of ten small ships are lost, or accidents occur, in the dark. Better a day late, than ———.

3. **Don't ever assume that you have passed a danger that you have not seen!** Get a lookout aloft and try to find it! Know you have passed each danger, before you cross it off and relax! It is a good idea to circle in **red** on your chart, and **number,** each **danger,** before you approach it. Then, methodically, in daylight, one by one, check them off!

4. **Don't delay, or hesitate, in any action toward safety.** If you wonder if you should shorten sail, you should shorten sail. If you so much as think you would be safer swung to two anchors, kedge out the second hook now! Most sails are blown because of a five minute hesitation.

5. **Don't,** for any length of time, particularly overnight, **leave a ship at anchor with no one aboard!** This is simply asking for it!

6. **Have a watch, awake and topside, at all times!** Night and day! And wearing a safety harness! How many small ships have been lost, simply because all hands were asleep below! Hard to believe, but true! (One of the great dangers of self-steering wind-vanes, if the wind shifts, but a few moments, unnoticed.)

7. **Don't rely for safety on Radios, Direction Finders, Radar, Loran, Consul, etc.** The only safe and sure way to know where you are is by continuous, thorough Celestial Navigation, backed up by careful, painstaking Dead Reckoning. Too often, it is the gadget ship that is lost, or in trouble.

8. **Don't tackle difficult or dangerous maneuvers, when you are exhausted.** Heave-to and sleep for a few hours. Exhaustion is the cause of most misjudgments, errors, and avoidable accidents at sea. In general, it is wise for the skipper to hold himself, within reason, in rested reserve before an impending danger.

9. **Don't go to sea, unless fully qualified, equipped and prepared!** Know and prove your ability to navigate on land. Have all proper safety and survival equipment on board, in the way of rafts, throw-over lights and markers, flares and dyes, and all in useable condition. Have ample food and water for twice the expected duration of the voyage. Above all, by your navigation, at all times know exactly where you are!

10. For the long hauls, the difficult passages, **don't take passengers, or crew, who never have been to sea before!** The risks involved, even with those who know what they are getting in for, are great enough. With those who simply imagine they will like it — the odds of success are hopeless! In general, stay away from

crew. Family are best! Girls, except for that rare one like Harry, are better crew than boys. They are not above helping in the galley. They bathe. They tend to keep their bunks and quarters neat. They will listen. They will take orders. In most cases, the male crew comes aboard to tell, and are somewhat aghast when told.

11. Don't fail to have a crash list, in order of priority, of things to do and take, **in case of abandoning ship in an emergency.** As, Sail, Oars and Mast, Sun-cover, Bailer, Flares, Signal Mirror, Water, Food, Sextant and Clock, Chart and 249 Book, Fishing Tackle, Ship's Papers, Passport and Money, etc.

12. Never abandon ship until she sinks out from under you!

* * *

And now, I've mentioned in this book about navigating a course line with only one Sight a day. I think it's time I got around to explaining it.

We are all familiar with the theory of an aeroplane flying on instruments by homing in on a radio beam. Actually, what the plane is doing is flying down a path in the air, much like a carpet of direction, at the end of which is the homing beacon. Actually, this is not a very different principle from our old Running Down a Latitude, with a daily Noon Sight, in the days of sail ships. The difference is, that the plane can do it in any direction, whereas the sail ship could only home in on a due east or west course.

But it is interesting to remember that the sail ship had only one chance, or one Sight, a day, to do this, yet that was enough. Essentially, what each is doing, the plane or the ship, is not caring too exactly how far along the beam they are, as long as they know they are on the beam. Realizing, as they each approach their homing point, they will have ample time and warning to realize they are coming in and, by local bearings and observations, make a safe landing.

Therefore, if we imagine our Course Line (our Rhumb Line) in traversing any substantial distance at sea from anywhere to anywhere, as really a homing beam, which we are following down — all that is important is that we stay on that beam and eventually we will get there! Any open-water Course Line has a constant Compass direction, providing we stay on it. And, as every Line of Position (LOP) from a Celestial Body means . . . a line somewhere on which we are, we have only each day (it can happen only once each day) to force the timing of an observation of the

Sun, so that the LOP will Plot on the chart exactly parallel to our Course Line.

If the two are exactly parallel, either they will fall one on top of the other, if we are Dead on Course, or the LOP will fall to one side or the other of the Course Line. Indicating a slight correction is necessary, over the next twenty-four hours, either to Starboard or Port, estimating the amount of correction so that, at the next sight the ship will again be back on course. This becomes an absorbing game, somewhat in the nature of marksmanship — to see how nearly, in that next twenty-four hours, you can bring her back, without crossing over, or falling short.

And so you proceed, from day to day, shooting the same Sight, the same sextant angle, at very nearly the same time. Each day producing that one parallel Position Line to the Course Line. Holding the ship, as closely as possible, sailing straight down that line. Until the Taff Log and our Dead Reckoning, checked by our occasional Latitude Sights, tells us we are getting near our destination. Then naturally we go into the full act. Dusk and Dawn Star fixes, combined with numerous Running Fixes from the Sun to bring us safely in.

If all the above is clear, logical, and acceptable, all we have to do is to figure out **how** and **when** to shoot the **Sun** to produce a **Line** of **Position** each day exactly parallel to our **Course Line.** And this is really quite easy. First, you will agree that, regardless of the Compass Heading of our Course Line, at some moment during the day the Sun's shadow must cross it at right angles. Heading NE or SW, it will occur somewhere near mid-morning; on a heading of SE or NW, it will happen during the afternoon. But to pin it down **exactly,** as to **Time** and **Sextant Angle (Hs)** and everything else, we need only our parallel rule, our right triangle, and our **249 Book, Vol. II.**

What we are going to do is to **start** near the **end** of that Sight's requirements — **with the exact necessary Azimuth** — and, so to speak, work the Sight **backwards,** till we determine the necessary **Sextant Angle,** to which we will fit the **Time.** For, as you know now, the Azimuth of every Sight is always exactly perpendicular, or at right angles, to the Line of Position, coinciding with our Course Line. So, with your parallel rule, walk the Course Line, whatever it is, over to the nearest Rose on the chart, and with it dead across the center of the Rose, set your right triangle against it on the center point of the Rose, and read off the exact 90° bearing to the Course Line, naturally taking it on the south or

Sun-pointing side, or what your Azimuth will have to be to give you that line.

Now what do we do?

We open **Vol. II** of the **249 Sight Reduction Tables,** to the correct **Latitude** page (same or contrary) for our **Position** at the beginning of this set-up, and taking the **Full Degree** of the **Sun's Declination** for that date from the Almanac, to give us the proper vertical column on that Latitude page, we run down that column noting the **Azimuths,** until we find the **Azimuth, exact** to that we have just taken from the **Compass Rose** on the chart. And what do we find beside it? The necessary **Hc,** or **Computed Height,** to produce that Azimuth, for those conditions of **Latitude** and **Declination.** So, our Hs, or Apparent Sextant Angle, **after** correction, should in theory, be very close to this Hc, though the Hc has not yet been corrected for **Main** or **Dec.** Using the Hc as a substitute value, extract the Main and Dip corrections from the Tables inside the Front Cover, and naturally **subtract** this from the Hc, for your estimate for a first try at hitting the **exact** Sextant Angle which we need to produce our Parallel Line of Position. Actually, you should attempt to estimate the "d" value also.

We are now set for the actual trial shooting. Adjust your Sextant to this angle, and at the Time of day roughly (depending on the Hc and angle your Course Line crosses the chart) that the Sun shadow will bear abeam of your ship, or Course Line, get topsides and follow the Sun up (or down in the afternoon, as the case may be) until, without altering your sextant, the Sun sits right on the Horizon and Time it! (The first attempt you might set up to shoot three sights, one figured a few minutes early, one for your Dead On estimate, the last a few minutes later than you figure, very slightly, altering the sextant accordingly. Then, commencing with the middle one, work and plot and select the right one.)

Well, you've about got it cornered. Work the Sight up, as you would any normal Sun Sight, and as you enter the 249 Book, you will see it zero right in on that column and Azimuth which you had selected. You'll be lucky if, in plotting the first Line, it falls exactly parallel, as naturally it will be thrown off a bit by the inclusion of the Minutes of Declination, which you probably didn't allow for in this first try. But this first Plotting will at once show you which way you are off. One way or the other, you should have shot a few seconds, or minutes, earlier or later. If later — you might still get up there for another crack; otherwise, you'll have to be patient until tomorrow.

Mind you, on those long monotonous days, this should be treated more as a game than a lazy man's way of navigating. But it will amaze you what a help it becomes in telling you where you are quickly and with enormous exactness in relation to your Course Line. I use it every day, regardless of how much additional curiosity I expend to supplement it. You will likely find that it is the "Shot" each day that packs the most excitement. With a couple of cocky helmsmen aboard, it is always good for a bet of a can of warm beer.

And so it goes. When you get it nailed down exactly, each day at about the same time you shoot that same Sight. Of course, if a blow, or something (Devil forbid that someone falls asleep at the helm for his entire watch) throws you way off, it is usually not worth trying to haul back to that original Course Line. Simply strike a new Rhumb Line from where you find you are to your destination, and repeat the whole above titillating process over again — secretly grateful, probably, that again you have something to do.

There is just one more little wrinkle, that you have to keep in mind. As you recall, the Sun travels from East to West 15° in each hour, or 1° during each four minutes of time. Which boils down to two minutes for each half a degree, or thirty miles, *et cetera* — all of which naturally is **added** or **subtracted,** depending on whether your course is taking you, each day, more to the East or more to the West. By your Dead Reckoning progress along your Course Line, regardless of how slightly your Longitude changes, each day, for this sight, you must make this very small adjustment in your head for the Time — this only so that you won't get up there **too late.** Too early is better. Actually, with the angle of your sextant remaining the same, the variation in Time will take care of itself, given a minute or two of early leeway — but you must **be there.** Unfortunately, the Sun rarely waits.

That's it!

And now, a fast word about **Great Circle sailing** and what it means. On long voyages in high Latitudes above 30° N or S (the higher the Latitude the greater the benefit, for, in Equatorial Latitudes, the chart distortion is negligible) on headings predominately East or West, as English Channel to New York, sailing a Great Circle course instead of a straight Rhumb Line can save you hundreds of miles. However, the more the heading approaches full North or South, the less the saving, as all Longitude Meridians are, themselves, Great Circle lines.

Contrary to the impression conveyed by every chart, the surface of the Earth is not flat, but round, being naturally a portion of a sphere. Therefore, confusing as it may at first appear, any **straight line** (other than exact N-S meridians) traversing the actual Earth, as being the **shortest** distance from any one point on its **convex** surface to any other, when **flattened** out on a chart, becomes a **curved** line. And conversely, any straight line drawn on a chart, when sailed, becomes a curved course on the sea. Hence, on a long voyage, to follow a straight line on a chart is usually **not** the shortest way from here to there. And so it pays to plot a curved line on the chart, to sail a straight line across an ocean.

The quick, easy method to determine your initial leg to start sailing any **Great Circle** Course Line, **plus** finding the total **Great Circle miles** from any point in the world to any other, is to let your **249, Vol. II** do the work for you in a matter of minutes, exactly as for any sight, by applying the following substitute formula to arrive at Hc for total distance and Z for first leg heading of your progressively changing curved Great Circle Course Line. The Hc arrived at must be subtracted from 90° and the remainder reduced **in Total** to **Minutes** (′) (degrees × 60, plus minutes) for your exact Great Circle distance in nautical miles.

As your advancing DR positions indicate that your present compass heading requires correcting to remain on the curved Great Circle Course Line, at each such point you simply begin new again, repeating the above procedure to determine the next heading, each such DR position then becoming your new point of departure. **Formula:** In 249 Book, Latitude Page becomes Latitude of Departure; LHA becomes Difference in Longitudes; Dec. becomes Latitude of Destination and thus cannot be above Lat. 30° N or S. Then Z equals **initial** Great Circle course heading. 90° minus Hc in Minutes equals total Great Circle Distance in Nautical Miles.

CHAPTER IX

A Word About Gadgets and Tools

THERE IS NOT MUCH TO BE SAID about the Sextant, itself — except that some are a lot better than others. When you live with one of these machines night and day, dependent on it under very adverse conditions, over the years the difference in price between a good one and a cheap one is soon forgotten. Whereas, the ease and accuracy of daily use goes on forever.

It is one thing to shoot the Sun from the high, relatively-motionless bridge of a steamship. Then as those lucky navigators do, wander in to a spacious, well-lighted, air-conditioned Navigation Room and sit down at a large desk, on which a gin-and-tonic has been left without fear of it spilling. And with every modern mechanical assistance, including a buzzer to summon a stewardess to bring you a sandwich, quietly work up a sight. No doubt that navigator might get away with a cheap sextant.

But you, on your small rolling, pitching, wet little chip, wallowing across one ocean after another (lucky guy) — with the sails and spars and rigging always in the way — are not so lucky. I know of one Skipper who installed two pad-eyes in the scuppers of his foredeck, to which he secured lines going aft from a shoulder harness, against which he leans forward — outward above the bow — to get clear of the headsails and allow him to use both hands in shooting a sight. (What a temptation for the crew to creep forward, "dirks" in their teeth, and slash both lines simultaneously as he shouts, "Mark!")

The German Plath, in my opinion, is the best sextant, albeit one of the most expensive. It is flawlessly constructed, beautiful in every way even to its lovely box, and the difference in taking a sight with it and a lesser sextant, if due only to the extra large finding area in the eyepiece, is unbelievable. I crossed the Pacific using an old Navy sextant, holding the Plath in reserve. Until one night I became so frustrated trying to pick up stars, that — never having looked through it before — I took the Plath out for the first time. I have never again touched that Navy prayer-wheel, and have never since had difficulty finding any star.

The British make some beauties, seductively small. Clean, with lots of polished brass. You can also get by with a plastic model for less than $15.00, sold even by such high-collars as Abercrombie & Fitch. So you have a wide range of choice. But get a good one. Also, it is a very solid idea to have two sextants aboard, when you consider how easily that one false move could force you to let go of even a lump of gold like that, to keep from going ears-over-tin-kettle into the drink. Every sextant should be fitted with a lanyard to go around your neck.

The word Sextant, as you may know, means the sixth part of a circle. In using it, for once you can truthfuly say that "it is all done with mirrors." It is very simple in function. You have two mirrors. One fixed on the frame parallel to the line of zero altitude. The second mirror is attached to a moveable arm, parallel to that arm. When the Sun appears in your eye, you simply have it reflected in an angular bounce, from the first fixed mirror, to the second moveable mirror, and presto, the amount of movement of the arm from zero gives you the height angle.

Occasionally you should check your sextant for zero-zero accuracy, which only means that when both horizons (fixed and moveable) are in exact alignment, the sextant reads Zero. If for any reason it doesn't, and you are at sea, **don't monkey with the mirrors!** To correct this, without touching anything, there is a very simple rule. You will notice that the scale on the graduated Arc of Degrees always goes below the Zero line (reading minus, or less than zero). If, with the two horizons in line, your sextant reads **more** than Zero, you **subtract** the amount of error from your Hs angle; if **less** than Zero you **add** the amount it is off. Just remember . . . "On is Off, Off is On."

And now, somewhere in the beginning of this book I stuck my neck out and said I would explain the several reasons why you **always** have to correct each sextant Hs angle, or Apparent Height. You still do have to do it, even if I don't tell you. The Dip correction I did explain, is because of the impossibility of giving different tabulations for every conceivable height of bridge. So the Tables give it with your eye in the water. Each navigator then lifts his own wet eye up from the water to his own particular level of dryness.

The **Main** correction, for all Celestial Bodies, is necessary for several reasons. As the Sun and the Moon rise, or set, you have noticed they appear bigger, and the higher they climb the smaller they get. This is due to atmospheric distortion — seeing them through more atmosphere when low, than when overhead. Consequently, the only constant point of accuracy in taking any measure from our Planet of another body is to relate always to the center of each. A part of that **Main** correction computes this (computed from the known diameter of each body), correcting your **Horizon** to **Lower** or **Upper Limb Angle**, so that it amounts to the angle you **would** have read on your sextant, had you been in the **center** of a glass **Earth,** aiming at a bull's-eye in the **Dead Center** of the celestial target. This has to do with **Parallax** and **semi-diameter.** If this doesn't satisfy you, drop around some evening and we'll sharpen a pencil.

Two other distortions also are taken into consideration in the **Main** correction: **Irradiation** and **Mean Refraction.** Actually you are better off to forget them. Irradiation has to do with the apparent enlargement of a bright object when seen against a dark background (a practical reason why fat women should avoid lying on black sheets). Mean Refraction has to do with that broken look a spoon has in a glass half full of water; meaning that Celestial Bodies are never quite where they appear to be. The lower to the water a body is, at sea, the more this monkeys it up. So now that we both know that much at least about the causes of the Main correction, let's quickly pour a tot of grog and move on to another item.

About the Taff Rail Log? A great little instrument, that I have finally about given up trying to use. Except in rare short-run uses, in clean water, where they are a great help in measuring the distance between dangers — or spotting the right headland or cove — I have found that, due to seaweed primarily, as well as current, leeway and set, over the long hauls they aren't worth dragging. They are not necessary for the Dead Reckoning estimate for an Assumed Position. The big spinners do slow you down; and the little Walker Knotmaster spinners seem to be harder and harder to find. But if you can keep it clean and spinning, sure — use it!

I had a shark take the last one of these big windmills, about a foot long, and swallow it whole. When I got to Bermuda, I innocently priced a replacement. Believe it or not, translating the pound price into dollars, they wanted eighty dollars — just for the spinner. (Perhaps they spotted me, though I did try to speak with a "blimey" accent.) I hope the marlin fisherman off Florida, who catches that shark, realizes the fortune hidden under that bulge in his belly.

It is surprising how quickly you can accurately estimate the speed you are making, by looking at the bubbles passing your hull. You'll soon be expert enough to guess it within half a knot. But if you have a spare hundred and seventy-five dollars, and can find one of the small Walker Knotmasters, and you have the space to store its box, fine. I am strictly against cutting one more hole through the hull, for this fancy little spinner about the size of an earring, that you drag through the weeds, pulling it in-and-out through the hull to clean it — the shaft of which activates a tiny dynamo, which in turn spins the hands on several dials — all of which with the corrosion, gumming and seaweed, says you are going two knots, when even your wife knows you are doing five. But, suit yourself. I'm going to move on to something really important.

Your watch! Your Navigation Wrist Watch!

This watch is what you really live by. For me, there is only one, and I don't own stock in the company! The Rolex GMT Master, with the adjustable outer blue-and-red bezel for Full GMT Time, always indicated by the red 24-hour hand; waterproof, stainless steel, self-winding and with a Calendar!

The importance of a calendar watch does not sink in, until you sail well away from the Homeland. Until you become a moving speck among islands that do not care about dates, on a chain of great seas where, for months you do not meet another boat. Where on no day, for any month, in any language, on any broadcast is the date ever given, even on WWV. As long as you are continuously navigating, from day to day, your Almanac keeps fairly close track of the date for you. But comes a long storm, or that rest of long sleeps, that period of forgetfulness on the hook in some tropical atoll, and — suddenly you don't know what day it is! As far as navigation goes, you are through.

But that Rolex has a built-in calendar!

Should the above ever happen to you (luckily it never has to me) I have worried enough about it to try to figure out what you should do. Fortunately, this should hopefully occur only when in association with the land, at an island, a bay, some uninhabited cove. Otherwise, you would know the correct date from your daily navigating. So, guess the day you think it is, and try it. If your date is right, your LOP will hit right across that particular cove or island. If it misses, take the next day forward, then back — keep shooting sights for different days, until you hit the day that puts your LOP right where you are. Let's hope you never have to. You won't with a Rolex GMT Master.

The number of reasons why this watch is indispensable are all too obvious. With any other type of chronometer, you have that age-old problem of computing Greenwich Time — always on a twenty-four hour basis, from a clock face composed of twelve hours — which can be as much as twelve hours different from Greenwich. To unscramble the Midnights from the Noons, all shown as twelve, forward and backwards for East and West Longitude, is one of the great opportunities for error in any sight. A hazard so easy to avoid. Simply get a Rolex.

Next in importance, I believe, as far as tools go, is a good Fathometer, or depth indicator. Because of possible main battery failure, I prefer those powered by flashlight dry-cell batteries, which can always be replaced. A fathometer is absolutely essential for the safety and preservation of your ship — lead-lines to the contrary! When on shoals, a single position line from the Sun, combined with several soundings beneath your Course Line, in most cases gives a very accurate Fix, providing your chart gives sufficient detail as to the bottom.

And so, what about a Compass? If you are not stuck on that one that came with your ship, so small and so confused that you can't read the card a foot away from it, even if you could see through that blackened glass, then — if you don't already know about it — go look at a Kelvin-White (now Danforth-White, I believe) 7″ Constellation, with a 5° card. It has a beautiful magnifying domed lens, that you can read from anywhere in the cockpit, night or day. You should also have a hand-bearing compass, for taking bearings from headlands, lights, etc. as well as a spare ship's compass for extended cruising. I recommend also a small "tell-tale" compass over the Skipper's bunk.

And lastly, for a slight deviation from the tools of the navigator, though definitely related as far as the Course Line is concerned, there is the subject of self-steering devices. I am referring now only to the problem of the long-range cruising sail boat. The wind-vane has come very much into prominence in the last few years. Certainly it must do a reasonably good job under favorable conditions, or one would not see and hear so much of them. Yet, in my opinion, the best of wind-vanes is always a compromise — and I have been told so by men who have sailed the world with them. The greatest expert on the wind-vane, whom I have ever met, after completing a circumnavigation under sail without an engine, confided in me that he did not know whether he spent more time adjusting and tinkering with his vane, than he would have if he had just steered his ship by hand. In light airs, certainly in variables, they are definitely unreliable. Naturally, under engine, they are of no use at all. They are frightfully bulky, both when in use, or unstepped and on the deck. Yet, still, they are better than nothing.

There is only one perfect self-steering device for the cruising sail boat, which never can afford any large consumption of electricity. It is equally dependable in all combinations of wind force, wind direction, sail or engine. Reliable in every way as to holding one set course and one course only. This is a small hydraulic device powered by your free-wheeling propeller shaft, disconnected under sail by a clutch from engine and gearbox, rotated by the propeller turning due to the forward motion of your ship under sail. It operates with equal reliability under engine. It is compact, simple, trouble free. It is available, if you are lucky, from R. R. Vancil, in Vandemere, N.C. Of all the helps in long-range cruising, this to me is the ultimate.

CHAPTER X

Conclusion

WE'VE TALKED SOMEWHAT ABOUT THESE STARS, and joked a bit about the curious names that men have attached to them. As far as this book and our navigation are concerned, they are just so many distant points of winking light fixed in the night sky, from which, during countless "twilights" we will measure. Too often we don't give too much thought to any of them — even as we stare upward at their brilliant fascinating designs, through all the long lonely hours of the dark watches. And I suppose many will say that is the way it should be. Why tax our minds beyond what actually concerns us in the use of these stars? Who cares what keeps the fire going on the Sun, or prevents it from falling downward out of the sky?

But just on the chance that you find yourself lying in the moon-washed cockpit on one of those tropical nights — the music drifting out from the shore — and you happen for a moment to wonder about what those specks of light are that helped bring you to that lagoon, as certainly all men have wondered since first they stared upward from a tree or cave, then here are just a few disturbing facts. For actually no man can ever begin to grasp the awesome mystery of what you are beholding.

We hear the word, "Universe," and a great many of us imagine that reference is being made to those stars that we see each night, if we happen, in our haste, to glance upward. Looking at the stars we can see, we are no more seeing the "Universe" than a man, hearing thunder and aware of distant lightning, peers at the few scattered raindrops on his dark window pane and imagines he has beheld the storm. On a star-bright, moonless night, staring at the glittering array filling the heavens, we see only a few fiery bits of dying dust matter, so close to us that if the "known" universe were a dust bin the size of the Grand Canyon, we and every star we can see, could all be snuffed up into an ordinary-size vacuum cleaner hose in one and the same instant.

Imagine a giant wheel of flaming spokes above you high in the heavens, bigger than the whole night sky. This light-spangled wheel is spinning at an enormous speed and so the spokes tend to curve backward, thinning at the ends into brilliant tails. Such a wheel of thousands upon millions of stars is called a Galaxy. Between one galaxy and the next and the next is nothing, nothing — absolute lifeless nothing — only the coldest imaginable darkness. The galaxies, alone, are lighted, by myriad tiny candle flickers of burning, self-consuming stars, utterly remote from each other, as our small, lonely Sun. Our little solar system lies out on the farthest tip of such a flaming tail of a galaxy called the Milky Way Galaxy. And so, we see our own galaxy upright above us, seeing into its width from its outer edge. Far beyond our galaxy, over a million other galaxies have been seen with the most powerful telescopes. Then where, and how far away, is the next nearest galaxy?

It just so happens that you can see it, just barely, with your naked eye. You are looking out at it through all these close near bits of burning dust, which are every other point of light in the night that you can see. In your binoculars you will see a blurred spinning glow, like a small vague phosphorescent cocoon, which you must remember is so far out there in the night that it is as if all these other bodies in the sky were so close to you they almost were touching the lens of your glass — those rain drops on the window. In your glass find Alpheratz, that nearest bright Star South of the Chair of Cassiopia, that we mentioned back in the beginning of our talk about the Stars, as lying on the First Point of Aries. Now move your binoculars slowly in the direction of Cassiopia, about a third of that distance. Suddenly among those many faint Stars you will find this blur, which is the great Spiral Nebula, or Galaxy of Andromeda. This is a whole, complete and different galaxy, identical to our own Spiral Nebula, the Milky Way. It is the only other galaxy that you can see without a telescope, except possibly for the dim smudge of another behind the Constellation of Orion.

But let's confine ourselves just to our little Solar System and its neighbors. Then perhaps if we can get some vague idea of the size of the Sun and its relation to one or two of these relatively near, local Stars, maybe we can begin to remotely grasp the enormity of a galaxy such as Andromeda, or the Milky Way. How can we describe the size of the Sun, other than in millions of miles, so we have a chance of trying to imagine such vastness inside our small brains?

Lying in your cockpit, gaze up at the Moon, if it is there. Try to imagine that our Sun, in its full size, is a hollow ball, and that our Earth in its full size is suspended in the exact center of this great hollow shell, our Sun. Then if you can imagine that our Moon too, spinning around us as it does, and as far away from us as you see it there, is also in its full-scale relation to our Earth inside the Sun — that Moon is just halfway out to the circumference crust of the Sun. And our Sun is a little Star. If it were out where those nearest Stars are that you see, it would be so dim you might not see it at all. I'll tell you roughly just how small it is.

You remember Betelgeuse (old Beetle-Juice), that red Navigation Star forming the upper shoulder of Orion? Incidentally, it is red because it is relatively cool, this being so of all red celestial bodies, as Mars and Antares. Well, just for kicks, how big do you think old Betelgeuse is, out there? Small as it looks, would you guess it is bigger than our Sun? It is so big, that if you imagined our Sun now, in its full size, suspended in the center of Betelgeuse, our Earth, with its Moon, could travel its yearly elliptical course around our Sun — all inside of Betelgeuse! So when you aim your sextant at Betelgeuse and you have a hard time finding it on the horizon — it is not because you haven't got enough Beetle Juice. But our Sun is a much more solid little Star, sitting out there minding its own business of burning itself up into a dead clinker, as someday it will be. For if a piece of the Sun the size of a match-head fell on your foot, it would feel as if you had been hit by an iron cannonball the size of a watermelon, it is so heavy (though Sirius, the Dog Star, is twice as heavy and twice as hot); whereas Betelgeuse, for all its size, is hardly any heavier than our air. Just a great big balloon full of red air — without the balloon.

For a couple of more head-scratchers, try these. Light, coming from our Sun, takes eight minutes to reach us. Or when we squint at all that fire, we see it as it was eight minutes ago. By the same token, when you peer through your binoculars at that Spiral Nebula of Andromeda, that light you are seeing left that galaxy 900,000 years ago — long before the first forked-shape even resembling man had begun to totter around down here, beating his chest, and palming his way from tree to tree to keep from prat-falling.

And as to numbers, it may or may not give you comfort to know that there are getting to be enough of us now, cozily en-sconced on the dry patches of our wet little planet — some three billion — that we about equal the number of Stars in our Milky Way galaxy, that white misty band crossing the sky, which is the thickness of our galaxy, as if you were a bug inside the rim of a wheel looking upward at the hub. So take your pick. One of those Stars is yours.

Which is another way of saying that there are getting to be a great many of us around, while the combination of time and bio-logical urge are producing more every minute. So, if it is in any way possible for you to achieve that rarest adventure of momen-tarily leaving this crowded dryness, even if briefly — to sail away into some sea, there you will find every aspect of this Planet unchanged. You will find our Earth exactly as it was in the pri-mordial beginning of man, when first in a fragile hollowed log he satisfied his yearning to find his way, with the wind and stars, to some distant, new, unexpected mystery.

You will find the Trade Winds and the waves, the strange-tailed soaring birds and the brilliant fish, the green clouds and the gentle colors of the sky, exactly as they were in this remote time, as well as that first-seen dark tree-top-flatness of countless coral islands, still unfouled. For the moment still unscarred. But particularly you will find the mood and the spell of the Stars and the Planets, their gigantic mystery and their wonder, unchanged since the dimmest beginning of time. You will find the Sea the last remaining challenge. An identical challenge, in no way al-tered, to that which lit the eyes of Ferdinand Magellan, Christo-pher Columbus, Vasco Da Gama, Cook, Bougainville and others. So gamble, rip up the roots — take that chance!

For an instant in long time, men will have existed. During an inexplicable meaningless moment they will have appeared, peered about, disappeared. A vertical rising and instantly descending line on a graph in the indifferent eons of time. Nothing more. So, if you can, wring the most out of that nothingness! And so, by way of closing this small book, let me wish you "Soldier Winds." Quickly now — get on your way! Start exploring, navigating this small shrinking, life-saturated, indefinably wonderful only Planet that we will ever know.

We are presumed to have evolved from the upper reaches of the lower of two seas. Then oddly we conditioned ourselves to leaving the first sea of water, to enter the lower levels of the upper sea of gas, which now we inhabit. So what more natural place for you, or I, to be, than just at the gaseous bottom of one sea, floating on the sun-lit surface of the other, drifting with the winds in some tiny bark, guided by these fiery myriad friendly bodies wheeling overhead, bending them to our will and needs by the use of an infinitesimal object close at hand, called a sextant.

NAVIGATION ABBREVIATIONS AND SYMBOLS

The following abbreviations and symbols are only those used throughout this book.

Aries	♈
Assumed Position	AP
Away	A
Azimuth (360° compass)	Zn
Azimuth (180° E or W)	Z
Declination	Dec.
Interpolation value	d
Dead Reckoning	DR
East	E
Geographical Position	GP
Greenwich Hour Angle	GHA
Greenwich Mean Time	GMT
Latitude	Lat.
Local Hour Angle	LHA
Computed (tables) Alt.	Hc
Observed (corrected) Alt.	Ho
Sextant Altitude	Hs
Sidereal Hour Angle	SHA
Toward	T
West	W
Interpolation value GHA	v

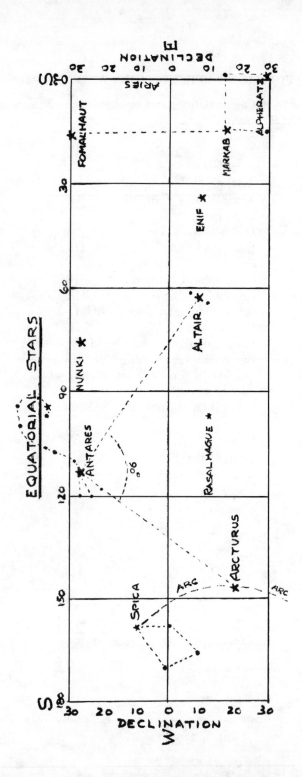

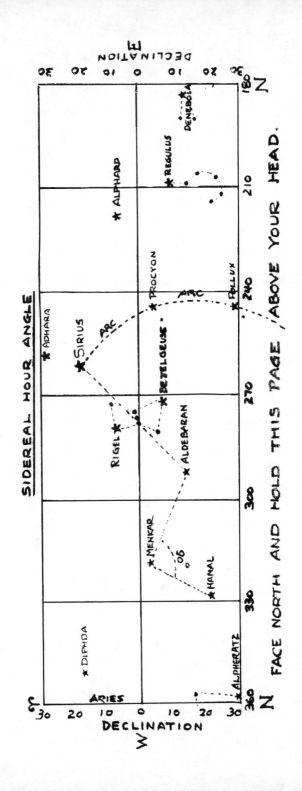

NOTES:

AUTHOR'S DEFINITIONS
OF A FEW YACHTING MISNOMERS

YACHT — The most expensive way yet discovered to go third class

JIBE — An (unintentional) maneuver that gave the boom its name

ANCHOR — A device for fastening yourself back on the planet, when you don't want to get on it all the way

FREE-BOARD — What most crew sign on for

BULLHORN — A device used by the skipper for lowering his voice

GAFF — Spare boom carried aloft

BOWSPRIT — A helpful device, if you signed on too many crew

OVERHANG — That extra part of the hull designed to float on air

DOUBLE NIGHT WATCH — Period of incessant talk

BAGGY-WRINKLE — The skipper's lower eyelids the third day of a storm

DALMATION HALF-TANGLE — The knot most heaving lines assume in mid-air

COMPANIONWAY — Where every companion aboard seems to congregate

HEAVE-TO — Second occurrence of seasickness

DOWN GUYS — Crew who won't get out of their bunks

DEAD RECKONING — Logical conclusion of seasickness on second day without respiration

COME ABOUT — Phrase usually preceding an exaggerated estimation of day's run

CREW — Those wonderful people who you take with you and who you can't wait to fly home at your expense

PORTHOLE — Result of encountering a barnacled log on left side of the hull

COCKPIT — Depression in deck intended for chicken fights, now degraded primarily just for cocktails — and chicks

PULPIT — Approaching harbor, the source of much non-ecclesiastical blasphemy

CLEW — That corner of the sail that helps you remember the names of the other two corners

NAUTICAL MILE — A land mile swollen by getting wet

COCKPIT DODGER — The one dry guy after the big wave

DOGHOUSE — A shelter in which much growling is heard and from which an old sea dog emerges in bad weather

NOTES: